47 Topics in Current Chemistry

Fortschritte der chemischen Forschung

Stereochemistry I

In Memory of van't Hoff

Springer-Verlag
Berlin Heidelberg GmbH 1974

This series presents critical reviews of the present position and future trends in modern chemical research. It is addressed to all research and industrial chemists who wish to keep abreast of advances in their subject.

As a rule, contributions are specially commissioned. The editors and publishers will, however, always be pleased to receive suggestions and supplementary information. Papers are accepted for "Topics in Current Chemistry" in either German or English.

Any volume of the series may be purchased separately.

ISBN 978-3-662-15538-7   ISBN 978-3-540-37922-5 (eBook)
DOI 10.1007/978-3-540-37922-5

Library of Congress Catalog Card Number 51-5497.

On the centenary of the publication in September 1874 of the work that laid the foundations of stereochemistry: "La Chimie dans l'Espace" by van't Hoff

J. H. van't Hoff in 1904

Contents

Stereochemical Correspondence Among Molecular Propellers

Prof. Dr. Kurt Mislow*, Devens Gust, Dr. Paolo Finocchiaro** and Dr. Robert J. Boettcher***

Department of Chemistry, Princeton University, Princeton, New Jersey, USA

Contents

* To whom correspondence should be addressed.
** NATO Fellow, 1972—73, on leave of absence from the University of Catania, Catania, Italy.
*** NIH Postdoctoral Fellow, 1972—73.

K. Mislow, D. Gust, P. Finocchiaro, and R. J. Boettcher

1. Introduction

A screw propeller is "a device consisting of a central hub with two, three, or more radiating blades symmetrically placed and twisted so that each forms part of a helical surface...". [1] There exists a large variety of compounds whose molecular conformations may be loosely described in this fashion. Although such compounds often differ greatly in their constitution, they are related in their static and dynamic stereochemistry.

What are the structural elements which permit characterization of a molecule as a "propeller"? It is possible to formulate rigorous definitions, the most inclusive of which would be that a molecular propeller is any molecule whose unsubstituted skeleton is of D_n or C_n $(n > 1)$ symmetry. Such a definition, however, is so broad as to be useless for our purpose. The above dictionary definition is largely intuitive when applied to molecules, and is meant to be utilized merely as an aid to identifying similar stereochemical features in molecules embracing a wide variety of chemical structures. Thus, it must be recognized that there exists no sharp dividing line differentiating chiral molecules which can be regarded as propellers from those which cannot. The concept of a molecular propeller is meant to be a unifying theme rather than a rigid classification scheme, and, in the last analysis, whether a structure should be viewed as a molecular propeller, *i.e.*, whether attribution of a propeller shape is sensible, lies mainly in the eye of the beholder.

To be identified as a molecular propeller, a molecule must possess two or more subunits which can be considered as "blades" (*e.g.*, aryl rings or spirocyclic rings) radiating from an axis of rotation (propeller axis). Furthermore, each blade must be twisted in the same sense[a] so as to impart a helical conformation to the molecule. Note that it is the *sense* of twist, and not the angle of twist, that is important. A molecule whose propeller skeleton bears different substituents on each blade, and consequently is expected to have different angles of twist for each blade, is encompassed by this deliberately flexible definition. For example, the three-bladed propellers diphenyl-*p*-tolylborane and phenyl-*p*-tolyl-*1*-naphthylborane, molecules of C_2 and C_1 symmetry, respectively, both have the D_3 skeleton of triphenylborane when stripped of their substituents (*i.e.*, methyl and benzo groups) and disregarding minor variations in the torsional angles. Thus, the propeller axis need not coincide with a molecular symmetry axis.

---

[a] Structures in which the blades are not all twisted in the same sense (*e.g.*, 1,3,5-triphenylbenzene [2]) do not fall under our definition of a screw propeller.

A three-bladed propeller whose skeleton is of D_3 symmetry may alternatively be viewed as the composite of three distinct two-bladed propellers of the same helicity, opposite to that of the three-bladed propeller. For example, triphenylcarbenium ion contains, in addition to the primary three-bladed propeller axis (C_3), three secondary two-bladed propeller axes coincident with the local C_2 axes of the phenyl rings. However, stereoisomerism resulting from secondary propellers is subsumed by the primary propeller, and therefore isomerism in molecules of this type may always be completely analyzed in terms of the primary propeller. In the case of molecules (propellers) of D_2 symmetry, the choice of any of the three C_2 axes as the primary propeller axis suffices for a complete description of the system.

Some molecules may possess two or more *independent* propeller subunits. For instance, tetraphenylallene may be thought of as the composite of two separate and independent two-bladed propeller systems joined at the central carbon atom. Although these propeller subunits are necessarily chiral, the molecule as a whole need not be. Thus, tetraphenylallene may in principle exist in both *meso* and *dl* forms.

Several examples of molecular propellers with varying numbers of blades are shown below *(1a—1j)*.[b] This collection is by no means exhaustive and could easily be extended. The present review will discuss the major features of isomerism and isomerization in several important classes of molecular propellers and will examine in some detail the stereochemical relationships among various groups of constitutionally dissimilar propeller molecules.

2. Stereochemical Analysis of Molecular Propellers

2.1. Static Stereochemistry

Our discussion will focus on three-bladed propellers with particular emphasis on work done in these laboratories, although other classes of propeller molecules will also be briefly considered. The analysis of the static stereochemistry of three-bladed propellers will be exemplified by a discussion of isomerism in molecules which feature three aryl groups (Ar) bonded to a central atom (Z).[c] Such molecules may be divided

---

b) Structural information indicating the propeller conformation is available for biaryls (*e.g.*, *1a* [3]), tris-chelates (*e.g.*, *1h* [4]), [4.4.4]propellane (*1e*) [5], 1,2,3-triphenylcyclopropenium perchlorate (*1i*) [6], hexaphenylbenzene (*1j*) [7], tri-*o*-thymotide (*1d*) [8], and for molecules in which two or three aryl groups are attached to a central atom (*e.g.*, *1b*, *1f*, *1g*). [c]

c) A discussion of isomerism and isomerization in such systems is available. [9]

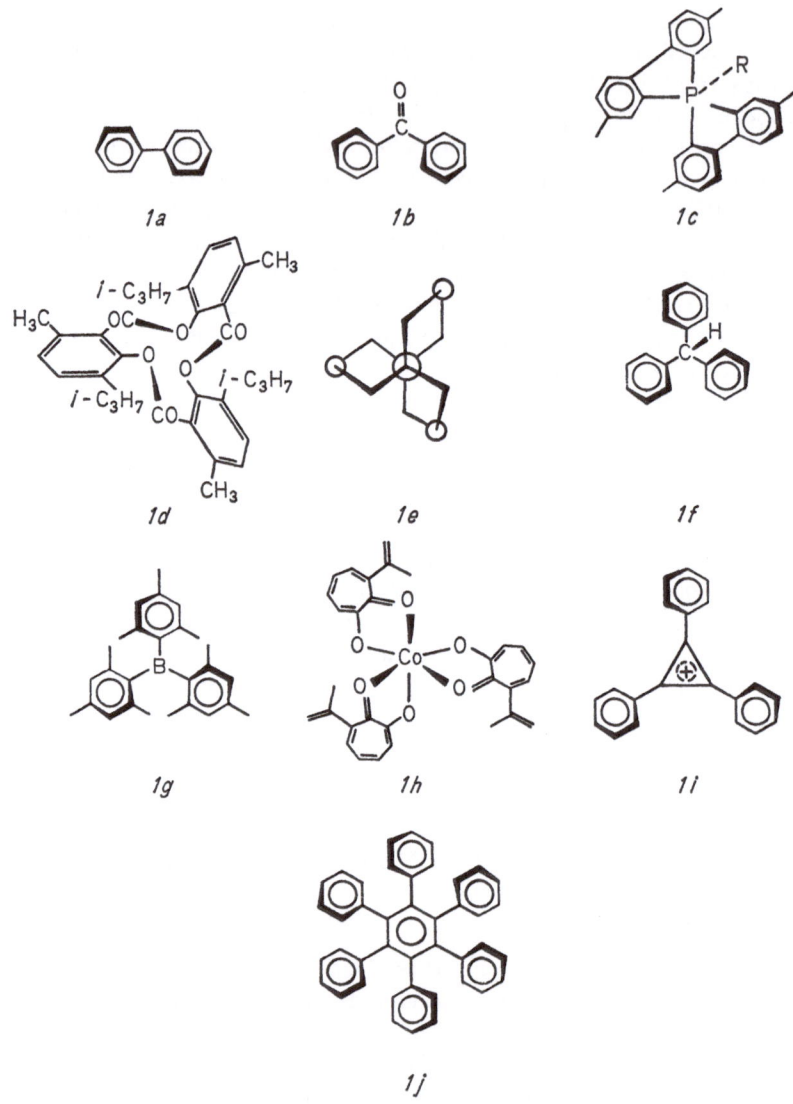

into two general classes: those whose skeleton is of D_3 symmetry (e.g., Ar_3Z molecules such as triphenylcarbenium ion), and those whose skeleton belongs to C_3 (e.g., Ar_3ZX molecules such as triphenylmethane). We will begin our discussion with molecules of the first type.

The conformation of Ar_3Z molecules is exemplified by the generalized structure shown in 2, which is depicted in one of several possible propeller forms. The three carbon atoms attached to Z define a plane (the *reference plane*), and the three aryl groups are twisted with respect to that plane.[d] In the general case, all three substituents (A, B, C) in *2* are different,

2

and none of the aryl groups has a local C_2 axis coincident with the Z—C bond.

In *2*, all three substituents may lie on the same side of the reference plane; alternatively, A, B, or C may be on the opposite side from the other two. Isomerism resulting from these four arrangements is analogous to the more familiar *cis-trans* isomerism. Each of these arrangements is chiral, even in the absence of helicity (*e.g.*, even when all rings are perpendicular to the reference plane), and may exist in two enantiomeric forms which differ in configuration with respect to the reference plane. Thus this plane may be treated as a plane of chirality. This is illustrated in the top portion of Fig. 1 where the two enantiomeric structures each have all three substituents on the same side of the reference plane. There are consequently $4 \times 2 = 8$ stereoisomeric structures, *i.e.*, four *dl* pairs.

Since each of the structures in the top portion of Fig. 1 may be converted into a propeller whose sense of twist is that of either a right- or left-handed helix, each structure gives rise to two diastereomeric propellers. The two propellers differ only in their configuration with respect to an axis of chirality (helicity), a line passing through Z and perpendicular to the reference plane. This is illustrated by the central portion of Fig. 1 which depicts the four stereoisomers (two *dl*-pairs) generated from the enantiomers in the top portion by the described twist operation. Note that the helicity is independent of the configuration with respect to the plane of chirality. Since there are eight ways to arrange the substituents with respect to the reference plane, and since a given arrangement also exhibits helicity, there are thus $8 \times 2 = 16$ isomers (8 *dl*-pairs) possible for the general Ar$_3$Z system represented by *2*.

---

[d] The central atom Z need not lie in the reference plane, and is expected to do so only if the reference plane is also a plane of symmetry, or if there is at least one molecular C_2 axis coincident with a Z—C bond. For conformations such as that shown for *2*, neither of these conditions is fulfilled, and there is therefore no reason why Z should lie in the reference plane. [9]

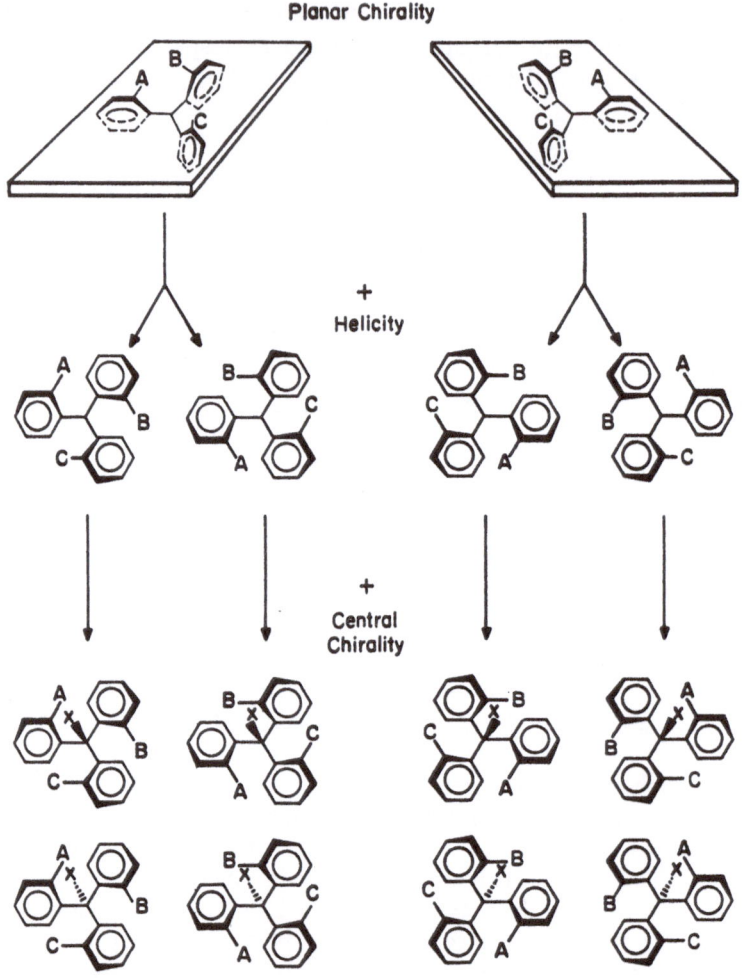

Planar Chirality

+
Helicity

+
Central
Chirality

Fig. 1

Degeneracies arise whenever two or more aryl groups are constitutionally identical, or whenever one or more of these groups has a local C_2 axis coincident with the Z—C bond. Table 1 lists the number of isomers for each possible degeneracy.

Molecules of the type Ar_3ZX may be represented by the generalized structure *3*, where X is any group with local conical symmetry on the time scale of observation. It is apparent that in addition to the elements of isomerism present in the Ar_3Z case, *3* possesses a center of chirality. Thus, each of the Ar_3Z structures shown in the central portion of Fig. 1

Table 1. Number of isomers for Ar₃Z systems

Number of identical rings	Number of rings with C_2 axes			
	0	1	2	3
0	16	8	4	2
2	8	6	2	2
3	4	0	0	2

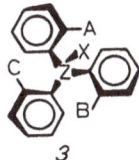

3

may be converted into two Ar₃ZX structures which differ only in their configuration with respect to the chiral center, as illustrated in the bottom section of the Figure. Since central chirality is independent of both planar chirality and helicity, there are thus $16 \times 2 = 32$ possible isomers (16 *dl*-pairs) in the general case represented by *3*. Degeneracies arise under the same conditions as in the Ar₃Z system, and Table 2 lists the number of isomers for each possible degeneracy.

Table 2. Number of isomers for Ar₃ZX systems

Number of identical rings	Number of rings with C_2 axes			
	0	1	2	3
0	32	16	8	4
2	16	8	4	2
3	8	0	0	2

Although the above analysis of isomerism in molecular propellers with three blades has been described in terms of molecules featuring three aryl groups bonded to a central atom, it is intuitively obvious that the analysis is equally applicable to other analogous three-bladed propeller molecules. Clearly, the points of analogy are independent of parameters such as the kind or number of atoms and bonds in each molecule; rather, they include such features as the symmetries of the

7

molecular skeletons and the symmetries and relationships of certain subunits of the molecular structure.

The question of isomerism is closely related to that of isomerization, and further elaboration of this point will follow a discussion of the dynamic stereochemistry of molecular propellers.

2.2. Dynamic Stereochemistry

Stereoisomerization in a molecule of the type Ar_3Z (2) may occur by rotation of the aryl groups about the Z—C bonds. In principle, there are numerous physical motions (mechanisms) which are capable of converting a given isomer into each of the 15 other isomers, or into itself. However, evidence to date for molecules of the type Ar_3Z is interpretable in terms of one of four general classes of mechanisms [9], called "flip" mechanisms, which were first suggested by Kurland et al., for isomerizations of triarylcarbenium ions.[10] In each of the four classes, zero, one, two, or all three aryl groups flip, i.e., rotate through planes perpendicular to the reference plane, while the remaining groups rotate through the reference plane (Fig. 2).[e] Each flip mechanism reverses the helicity of the reactant

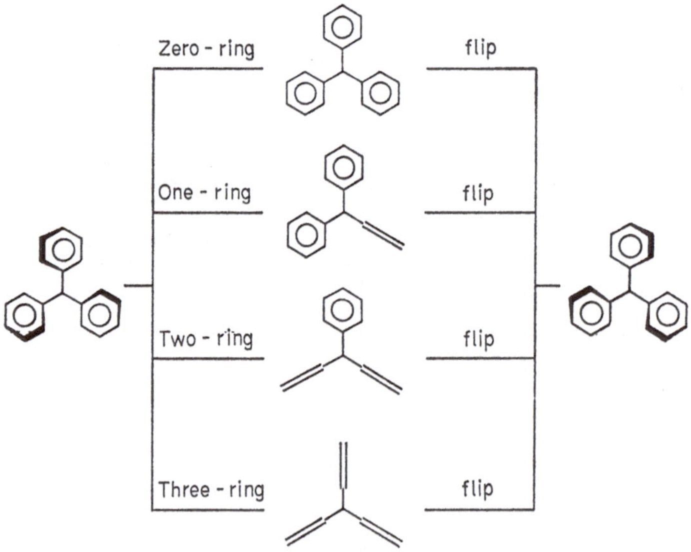

Fig. 2

---

[e] The transition states depicted in Fig. 2 are idealized and are meant to imply only the net result of the transformations.

molecule; in addition, in the general case represented by 2, each flip mechanism leads to a different stereoisomer. Thus, for a given isomer, eight single-step isomerization pathways are available by way of four classes of mechanisms: one zero-ring flip, one three-ring flip, three two-ring flips, and three one-ring flips (Fig. 3). For the entire set of 16 isomers there are therefore $(16 \times 8)/2 = 64$ interconversion pathways by way of these mechanisms.

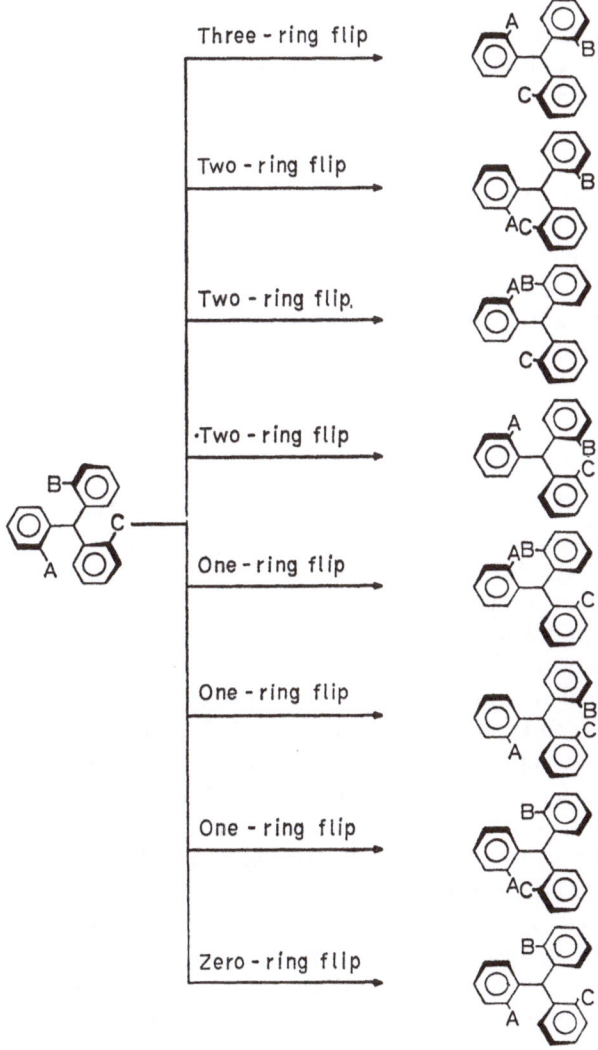

Three - ring flip

Two - ring flip

Two - ring flip.

·Two - ring flip

One - ring flip

One - ring flip

One - ring flip

Zero - ring flip

Fig. 3

The results of isomerizations by these mechanisms for the 16 isomers of the general Ar_3Z case have been completely tabulated elsewhere [9], and therefore we will here summarize only a few of the more salient results of this analysis. The zero-ring flip always results in enantiomerization, and is the only mechanism which achieves enantiomerization in one step. The three-ring flip, on the other hand, changes only the helicity of the reactant molecule. The two- and one-ring flips may affect all elements of isomerism, and a minimum number of three two-ring flips is necessary to achieve enantiomerization. In the general Ar_3Z case, enantiomerization solely by the one-ring flip pathways is impossible. Combinations of isomerizations by the various pathways are conceivable, and the results of such combinations are readily depicted by use of the appropriate graphs or matrices. [9]

Propellers of the type Ar_3ZX may be analyzed in an analogous manner. [9] The four classes of rotational mechanisms are retained, and an additional non-rotational pathway is added: inversion at Z along the Z—X bond. Such inversion changes the configuration only with respect to the center of chirality, and has no effect on the other elements of isomerism. Enantiomerization pathways for an Ar_3ZX molecule of type *3* thus must include an odd number of inversions.

3. Stereochemically Correspondent Molecular Propellers

We have noted that all three-bladed propellers with D_3 skeletons are in some ways structurally analogous. The analogy extends to their isomerization pathways. We will investigate the nature of this relationship by discussing the concept of stereochemical correspondence and then illustrating its application to propeller molecules.

It is generally recognized that in order to analyze the stereochemical features of real systems (molecules, or groups of molecules and reactions) it is useful to abstract those features which determine the properties of interest and to represent them in terms of a single model system. It is possible to perform an isomorphic mapping of the essential stereochemical features of each of the real systems onto this model, and an equivalence relation will thus exist between the real systems. We shall say that any two systems whose static stereochemistry (structure, conformation) and conformational dynamics may be represented by the same abstract model exhibit *stereochemical correspondence*.

The concept of stereochemical correspondence suggests that if one is willing to purge a given model of all non-essential chemical features, the resulting abstract model may well be applicable to a far wider variety of real chemical systems, and therefore has the potential of uniting hitherto disparate fields of inquiry.[f] A few of the more rewarding examples of the stereochemical correspondence among systems of propeller molecules will be discussed below.

We have previously noted [9] that three-bladed propeller molecules with D_3 skeletons exhibit stereochemical isomorphism with respect to stereoisomerism, and with respect to stereoisomerization when all pathways are considered. This fact is dramatically illustrated by the stereochemical correspondence between the Ar_3Z systems and the transition metal tris-chelates. We shall first describe a model in the sense discussed above which may be used to analyze the stereochemistry of one, and therefore of both of these systems. In general, such a model will consist of two components: a topographical reference structure, and a set of allowable operations on this structure. For example, let structure 4 represent the skeleton of the molecule, where the numbered vertices identify substituent sites connected in a pairwise fashion. The abstract model may then be defined as structure 4 together with the following operations: reflection of the structure as a whole, and/or permutations among the substituent sites which maintain the pairwise connectivity of the structure. This model or similar ones have been employed for tris-chelates [12–16] and for systems of the type Ar_3Z. [g],[9]

4

11

Obviously, all tris-chelates whose stereochemistry is analyzable in terms of the above model are stereochemically correspondent: similarly, compounds of the type Ar_3Z, such as triarylboranes and triarylcarbenium ions are also stereochemically correspondent. However, the more interesting correspondence is that between the class of tris-chelates and the class of Ar_3Z molecules. Although these two classes of molecules differ enormously in their chemical properties and reactions, the concept of stereochemical correspondence tells us that their stereochemical attributes are necessarily closely related. For example, a tris-chelate and a stereochemically correspondent triarylborane will each have the same number and kinds of stereoisomers, interconvertible by the same number and kinds of rearrangement pathways.

This last statement requires careful interpretation. In the first place, the abstract model described above provides no information concerning the stabilities of the conceivable ground state structures. In a real system, some of these structures may be energetically disfavored; chelate structures corresponding to conformationally stable triarylborane isomers may well be relatively unstable, and vice versa. In addition, the model describes isomerizations in terms of permutations and therefore the statement that two isomerizations are stereochemically correspondent signifies only that they interconvert stereochemically correspondent ground-state structures; *no information concerning the actual physical mechanism of the transformations is implied*. In order to avoid confusion, we will use the term "rearrangement" in this paper exclusively to denote the permutational consequences of a particular isomerization.[17] Thus, a two-ring flip *rearrangement* which converts molecule A to molecule B implies only the net result $A \rightarrow B$, whereas isomerization of A to B by the two-ring flip *mechanism* implies physical motions and the associated transition states, dynamics, and energetics.

The stereochemical correspondence between tris-chelates and Ar_3Z molecules is strikingly reflected in many of the properties of the two systems. For example, it was mentioned above that there are 16 conceivable rearrangements of the Ar_3Z molecule 2 which will convert the molecule into 15 different isomers, or into itself. Eight of these rearrangements are flip rearrangements, whereas the other eight involve no change of helicity. These 16 rearrangements form a mathematical group isomorphic to the abstract group $D_2 \times D_2$ under successive composition.[18] It necessarily follows that there is also a group of 16 conceivable rearrangements for a stereochemically correspondent tris-chelate system. The latter group has been discussed by Eaton and Eaton, who have stated that there are 16 operations (permutations and permutation-inversions) which "constitute the complete group of rearrangements of stereochemically non-rigid tris-chelate complexes." [14] These authors have applied their analysis to the dynamic nmr behavior of chelates of the type $M(A—B)_3$. It follows from the principle of stereochemical correspondence that the dynamic nmr behavior of stereochemically correspondent Ar_3Z compounds may be analyzed in *precisely* the same manner.

It must be emphasized that although the two groups of rearrangements mentioned above are stereochemically correspondent, the associated physical processes are vastly different in terms of mechanism and energetics. For example, the stereoisomerization mechanism of lowest energy for tris(α-isopropenyltropolonato)cobalt(III) (*1h*) has been discussed in terms of the trigonal twist mechanism.[4] Although the resulting rearrangement is stereochemically correspondent to the three-ring flip rearrangement, it is the two-ring flip mechanism which is of lowest energy for sterically crowded triarylboranes (see Section 4.1). We have also noted [9] that the two-ring flip rearrangement is stereochemically correspondent to the Rây-Dutt rearrangement of tris-chelates. Although the groups of rearrangements of tris-chelates and molecules of the type Ar_3Z differ in many of their details, the structures of the two groups are identical, *i.e.*, the two groups are abstractly equal.

The stereochemical correspondence between tris-chelates and Ar_3Z molecules is reflected in the topological representations of the results of isomerizations by the corresponding rearrangements in each system. For example, Fig. 4 consists of two nested cubes which are connected at corresponding vertices. As has been pointed out [9], if the vertices represent isomers of the general Ar_3Z system, then Fig. 4 is the graph of the two- and three-ring flip rearrangements. The edges defining each of the two cubes represent rearrangements by the two-ring flip, whereas the eight edges joining the cubes represent three-ring flips. By the same token, if the vertices of Fig. 4 represent the isomers of the tris-chelate system,

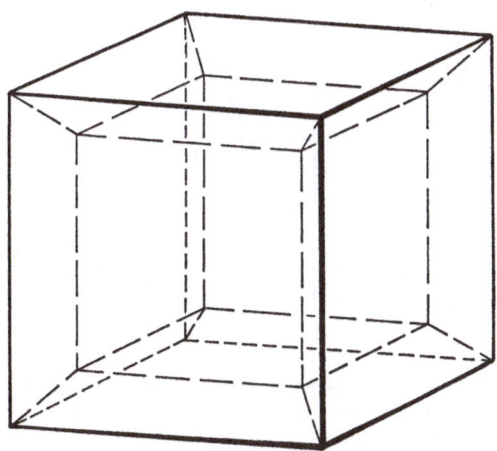

Fig. 4

then the Figure is the graph for the Rây-Dutt (edges defining each of the cubes) and trigonal twist (edges joining the cubes) rearrangements.[h],[12]

There are numerous other examples of stereochemical correspondence in propeller molecules. The propellane [19] *1e* has only one energetically reasonable isomerization mechanism available: a twisting motion about the C_3 axis which corresponds to the three-ring flip as well as to the trigonal twist rearrangement (Fig. 5). Similarly, only one isomerization mechanism (stereochemically correspondent in the permutational sense to the zero-ring flip) is energetically reasonable for tri-*o*-thymotide (*1d*). [8,20]

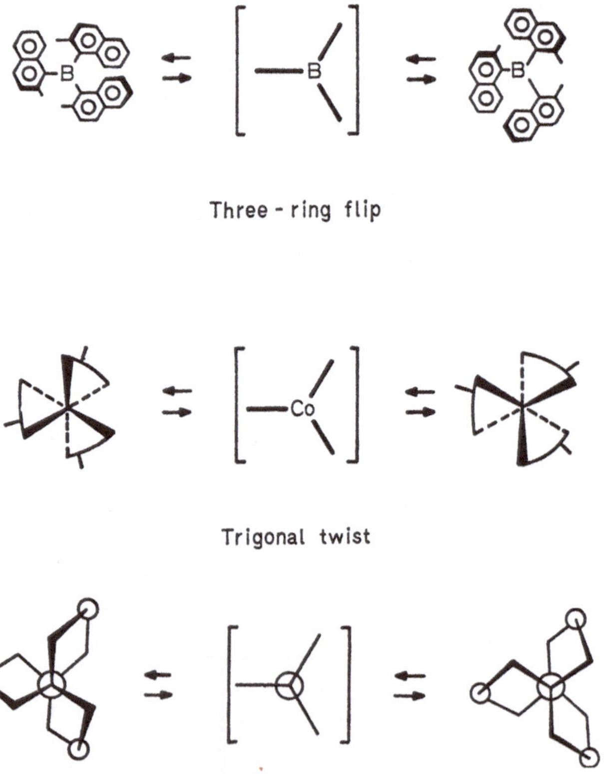

Three - ring flip

Trigonal twist

Fig. 5

Stereochemical correspondence between molecules of structurally disparate types may be further illustrated by a comparison of two

---

h) Since these classes of rearrangements reflect only the structure of the products and reactants, they may be equated with the "modes" of Musher [16], and with the corresponding classes of other workers. [13–15,17]

classes of two-bladed propeller molecules. These are diaryl molecules of type Ar_2ZX, *e.g.*, benzophenone *(1b)* or diphenylmethane, and spirocyclic pentacoordinate molecules, *e.g.*, phosphorane *1c*.[21]

Isomerizations of molecules of the type Ar_2ZX may be analyzed[9] in terms of flip mechanisms analogous to those described for three-bladed molecular propellers. The reference plane is now defined by the central atom Z and the two aryl carbons bonded to it. Thus, the two-ring flip involves rotation of both rings through planes perpendicular to the reference plane, the one-ring flip involves rotation of one ring through the reference plane and the other through a plane perpendicular to it, and a zero-ring flip consists of rotation of both rings through the reference plane. By contrast, isomerization in spirocyclic pentacoordinate compounds has conventionally been discussed in terms of the Berry pseudorotation mechanism[22], which involves pairwise interchange of ligands in the two apical sites of the idealized trigonal-bipyramidal ground-state structure with two of the three equatorial ligands by way of a square pyramidal transition state. The non-permuted ligand is called the pivot.

The stereochemical correspondence between molecular propellers of the type Ar_2ZX and spirocyclic pentacoordinate systems is most readily appreciated through a comparison of the graphs representing isomerizations in each system. Since graphical representations only show the *results* of isomerization by various mechanisms (*i.e.*, permutations), they actually depict rearrangement modes, rather than physical mechanisms. Thus, the Berry pseudorotation mechanism discussed above is representative of a mode M_1 rearrangement, in the nomenclature of Musher.[23] Other mechanisms belonging to the same mode, such as the turnstile rotation mechanism[24], are equally valid representatives of this mode. The mode M_1 rearrangements of a pentacoordinate molecule may be depicted topologically by the Desargues-Levi graph[25] (Fig. 6a). Each isomer (vertex) is designated by the indices of its apical ligands; thus, 14 has ligands 1 and 4 in apical positions, and $\overline{14}$ is its mirror image. The 30 edges of this graph denote mode M_1 rearrangements. In spirocyclic systems, four vertices of this graph may be deleted, since the rings cannot span the axial positions. The resulting subgraph (Fig. 6b) consists of 16 vertices and 20 edges. Additionally, a spirocyclic structure with a di-equatorial small ring is severely strained, and the vertices whose designations include the numeral 5 therefore represent high-energy intermediates and may thus also be eliminated, yielding the graph in Fig. 6c. In this realization, the vertical edges represent mode M_1 rearrangements, whereas the edges of the top and bottom faces of the cube represent rearrangements by mode M_4. From the above derivation of this graph, it is evident that an M_4 rearrangement may alternatively be expressed as two successive M_1 rearrangements involving a high-energy inter-mediate with a diequatorial ring.

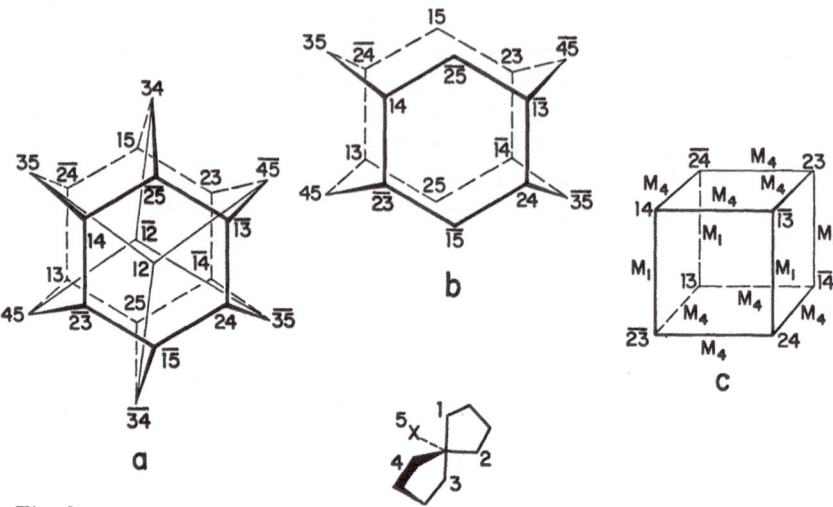

Fig. 6

Fig. 6c is isomorphic to the graph of the two-ring and one-ring flip rearrangements of an Ar_2ZX system in which the two aryl groups are constitutionally different, and neither of the two groups has a local C_2 axis [9] (Fig. 7). In this graph, the vertical edges (labeled AB) represent two-ring flip rearrangements, whereas the edges of the top and bottom faces of the cube (labeled A or B) denote one-ring flips. Thus, a mode M_1 rearrangement of a spirocyclic pentacoordinate molecule is stereochem-ically correspondent to the two-ring flip rearrangement, and a mode M_4 rearrangement is stereochemically correspondent to the one-ring flip rearrangement of Ar_2ZX molecules. The structures represented by the vertices of Fig. 6c and Fig. 7 are also stereochemically correspondent. In each graph, diagonally opposite corners of the cube represent enantio-mers, and any two adjacent isomers have opposite helicities.

It should be noted that the above discussion only partially exem-plifies the stereochemical correspondence between Ar_2ZX compounds and spirocyclic compounds, and is not a complete description of possible isomerization pathways. For example, the zero-ring flip rearrangement of Ar_2ZX molecules and the stereochemically correspondent mode M_3 rearrangement [23] of spirocyclic phosphoranes would be expressed graphically as edges joining diagonally opposite vertices of the cubes in Fig. 6c and Fig. 7. In all, there are eight conceivable rearrangements which will convert a given Ar_2ZX molecule (or spirocyclic phosphorane) into each of the seven other isomers, or into itself. These eight rearrange-ments form a mathematical group, isomorphic to the abstract group

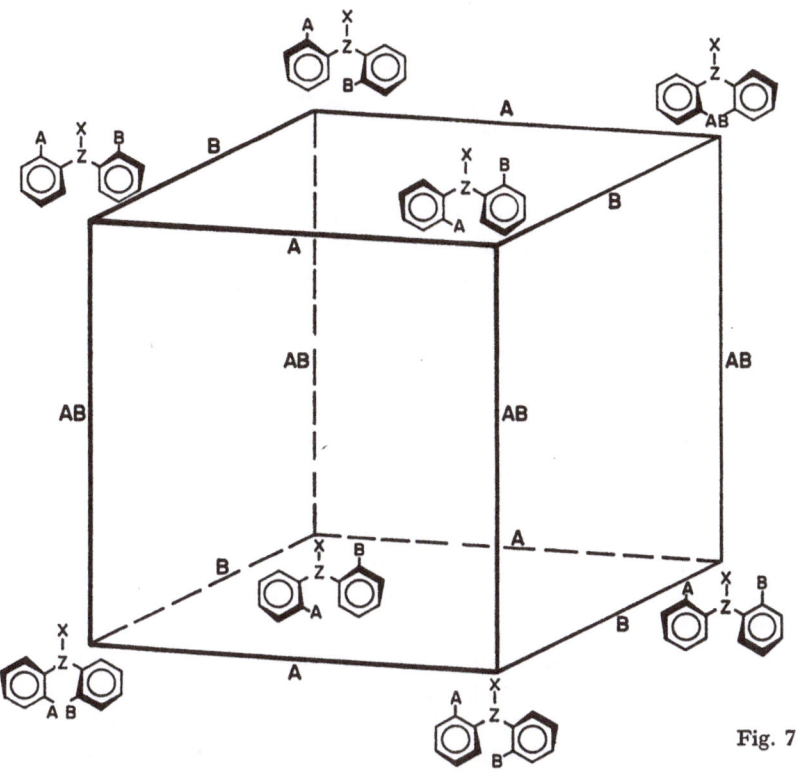

Fig. 7

$C_2 \times D_2$, [21)] and the group of rearrangements of the Ar_2ZX compounds and the corresponding group for the spirocyclic systems are abstractly equal.

Chemical applications of the analysis of spirocyclic molecules and of molecules of the type Ar_2ZX abound. For example, Hellwinkel [26)] has studied the [1]H-nmr behavior of molecule *1c*, where $R = CH_3$. At $-60\,°C$ the [1]H-nmr spectrum of the molecule features two aromatic methyl resonances. This result is consistent with a trigonal bipyramidal propeller conformation with two diastereotopic sets of methyl groups: two axial and two equatorial, as shown in *1c*. When the sample is warmed, these signals broaden and coalesce to a singlet at $-26\,°C$ ($\Delta G^{\ddagger}_{26}$ 12.5 kcal /mol). [26)] This coalescence was interpreted as arising from a rearrangement by mode M_1. In contrast to *1c*, the arsorane *5* has four diastereotopic pairs of methyl groups on the rings in the propeller (C_2) conformation. [27)] The [1]H-nmr spectrum of *5* at *ca.* 80 °C features only two resonances for these methyl groups. This observation is consistent with rapid isomerization by a mode M_1 rearrangement on the nmr time scale, since this process results in pairwise averaging of the four sets of

17

Me   Me   Me

Me—⟨O⟩⟨As⟩⟨O⟩—Me

Me—⟨O⟩⟨O⟩—Me

Me   Me

5

methyl groups.[27] The two observed resonances finally coalesce at 155 °C, and this process is interpretable in terms of a mode M_4 rearrangement, the equivalent, as was said above, of two consecutive Berry pseudorotations (mode M_1).[27] Similar studies have been carried out in stereochemically correspondent phosphorane systems.[28],[1]

In the Ar_2ZX systems, rearrangements corresponding to all modes of rearrangements of spirocyclic compounds are conceivable, but the relative energetics of the associated mechanisms are often quite different. Dimesitylmethane is stereochemically correspondent to *1c* ($R = CH_3$) and therefore two *ortho* methyl group ^1H-nmr resonances are expected for the molecule in the propeller conformation. However, only a single resonance is observed for these groups, even at *ca.* −95 °C.[31] In the absence of accidental isochrony, this result is consistent with several modes of rearrangement, including the two-ring flip and one-ring flip rearrangements. Empirical force field calculations[31] indicate that enantiomerization by the one-ring flip mechanism is the stereoisomerization pathway of lowest energy. Thus, stereoisomerization is presumed to occur by a one-ring flip rearrangement (stereochemically correspondent to a mode M_4 rearrangement of phosphoranes), in contrast to the phosphoranes discussed above where the mode M_1 rearrangement occurs with lowest energy.

We have seen that the concept of stereochemical correspondence is a useful analytical tool. If two or more systems are recognized to be stereochemically correspondent, an analysis of stereoisomerism and stereoisomerization in one system can be directly applied to the others, by way of the model abstracting the key features of both.

4. Selected Studies of Three-bladed Propellers

4.1. Triarylboranes

The stereochemistry of triarylboranes provides a typical example of isomerism and isomerization of molecular propellers with idealized

---

[1] Note that spirocyclic phosphoranes such as *1c* where $R = aryl$ [26,29,30] are stereochemically correspondent to three-bladed molecular propellers such as Ar_3Z compounds and tris-chelates, rather than to Ar_2ZX compounds.

D_3 symmetry. [32] This conformation is adopted by trimesitylborane (*1g*) in the solid state, and the low-temperature [1]H-nmr spectra of divers triarylboranes which have substituents other than hydrogen in all six *ortho* positions are consistent with this geometry in solution. At higher temperatures, the spectra of such molecules reflect rapid stereoisomerizations which are interpreted in terms of flip mechanisms.

Three - ring flip or
zero - ring flip

Two - ring flip or
one - ring flip

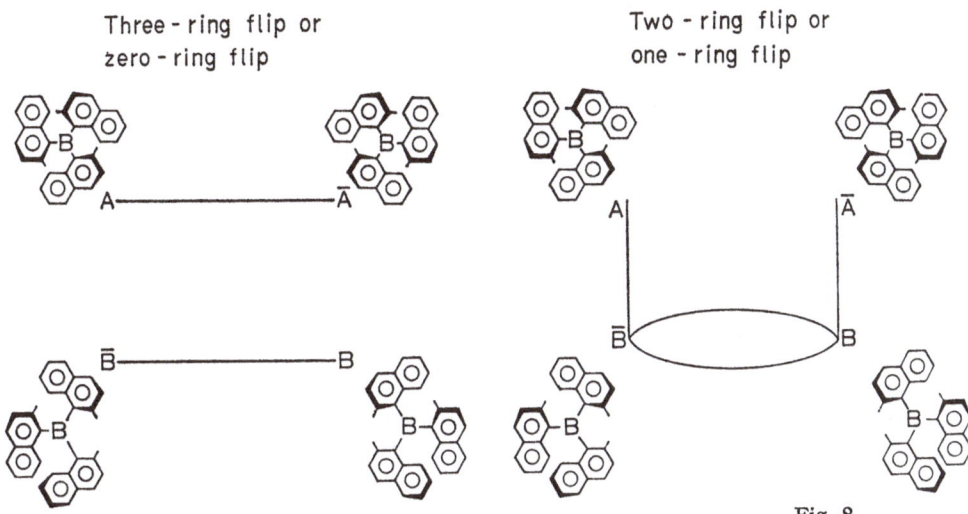

Fig. 8

One molecule of this type is tris-1-(2-methylnaphthyl)borane (*6*), [32] for which four isomeric propeller conformations are possible (Table 1). These isomers make up two diastereomeric *dl*-pairs (Fig. 8). One set of enantiomers (A and $\bar{A}$, hereafter A$\bar{A}$) are of C_3 symmetry and consequently each enantiomer has three equivalent methyl groups. The other two (B and $\bar{B}$, hereafter B$\bar{B}$) have C_1 symmetry and each enantiomer has three diastereotopic methyl groups. These facts are reflected in the [1]H-nmr spectrum of the methyl region at $-70\,°C$ (Fig. 9). The three resonances of equal intensity are those arising from B$\bar{B}$ whereas the more intense upfield singlet derives from the methyl groups of A$\bar{A}$. At this temperature, the ratio of B$\bar{B}$ to A$\bar{A}$ is 3.0 to 2.7. As the sample is warmed, the population of B$\bar{B}$ increases relative to that of A$\bar{A}$, indicating a positive entropy for the equilibrium A$\bar{A} \rightleftarrows$ B$\bar{B}$, and a crossover temperature ($\Delta G^0 = 0$) below which A$\bar{A}$ is more stable. A plot of ΔG_T° *vs* T yields, for the equilibrium A$\bar{A} \rightleftarrows$ B$\bar{B}$, ΔH^0 0.61 ± 0.05 kcal/mol and ΔS^0 3.1 ± 0.2 eu. The crossover temperature is therefore *ca.* $-76\,°C$. The major part of this entropy difference is accounted for by the differ ence in symmetry (C_3 *vs* C_1) of the two diastereomers ($R\ln 3 = 2.18$ eu, for the equilibrium as shown).

19

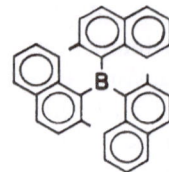

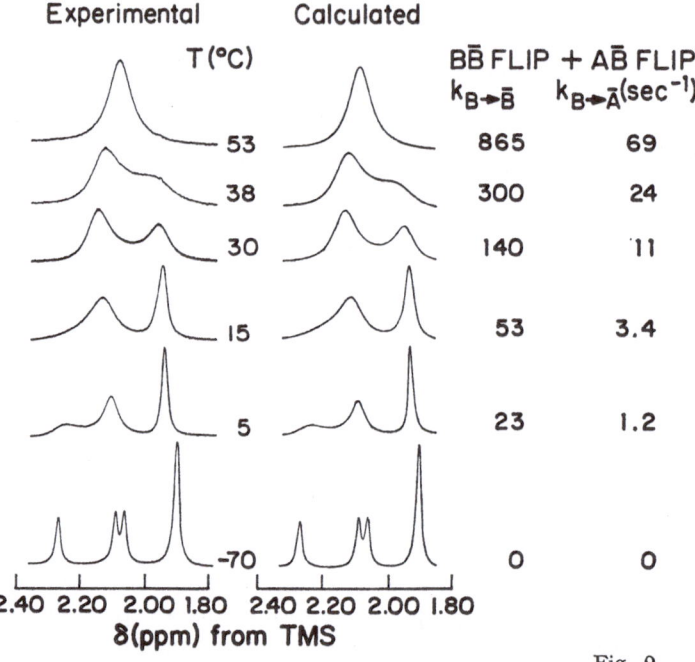

Fig. 9

When the sample of *6* is warmed to *ca.* 5 °C, the spectral lines broaden, indicating that stereoisomerization becomes rapid on the nmr time scale. The results of stereoisomerizations by the flip rearrangements are depicted graphically in Fig. 8. The graphs for all four classes of rearrangements are presented, although the zero-ring flip mechanism is doubtless of high energy because of the presence of large *ortho* substituents. The zero- and three-ring flips only result in enantiomerization, whereas the one- or two-ring flips may cause enantiomerization or diastereomerization, depending upon which of the pathways is followed. Fig. 9 shows that at *ca.* 15 °C the three methyl resonances of BB̄ have coalesced to a broad singlet, whereas the singlet for AĀ is virtually unaffected. This coalescence reflects enantiomerization of B and B̄. A detailed consideration [32] of the rearrangements depicted in in Fig. 8 reveals that rapid enantiomerization by the zero-ring flip would not result in any coalescence of signals, whereas the three-ring flip or either of the two diastereomeric one-ring flips interconverting B and $\bar{B}$ would cause the coalescence of only two of the three methyl group resonances of B$\bar{B}$. In the absence of fortuitously equivalent rates for two processes, e.g., the diastereomeric one-ring flips, the enantiomeric two-ring flips which interconvert B and $\bar{B}$ (the B$\bar{B}$-flips) are the only flip rearrangements which result in coalescence of all three methyl resonances of B$\bar{B}$, and thus are the only flip rearrangements consistent with the observed spectra at 15 °C. The two-ring flip is therefore the favored rearrangement pathway for enantiomerization of B$\bar{B}$.

When the sample is warmed to above 50 °C, all four methyl signals coalesce to a singlet (Fig. 9). This behavior indicates diastereomerization. The zero- and three-ring flips are incapable of accounting for this result. Fig. 8 shows that there are a pair of enantiomeric two-ring flips each of which interconvert either A and $\bar{B}$ or $\bar{A}$ and B. These rearrangements (the A$\bar{B}$-flips) result in the observed coalescence of all four methyl resonances to a singlet. Thus, both the observed stereoisomerizations of 6 (enantiomerization of B$\bar{B}$ and diastereomerization of A$\bar{A}$ and B$\bar{B}$) are interpretable in terms of two-ring flip rearrangements.

Quantitatively, line-shape analysis was used to determine rate data for these stereoisomerizations in terms of the two-ring flip mechanism. [32] The associated free energies of activation for the various exchange processes at 20 °C are shown schematically in Fig. 10. For the equilibrium B$\bar{B}$⇌A$\bar{A}$, ΔG°_{20} is 0.3 kcal/mol. For the conversion of B$\bar{B}$ to A$\bar{A}$, the calculation yielded $\Delta G^{\ddagger}_{20}$ 16.2 kcal/mol. and for the reverse reaction (A$\bar{A}$→B$\bar{B}$), $\Delta G^{\ddagger}_{20}$ 15.9 kcal/mol. The barrier to enantiomerization of B and $\bar{B}$ is $\Delta G^{\ddagger}_{20}$ 14.6 kcal/mol. Thus, at 20 °C the enantiomerization of B and $\bar{B}$ is energetically more favorable by 1.6 kcal/mol than that of A and $\bar{A}$.

Variable temperature ^{1}H-nmr studies of other hindered (in both ortho positions) triarylboranes have revealed [32] that these compounds are also mixtures of propeller-like stereoisomers on the nmr time scale at low temperatures, and that interconversion of these stereoisomers occurs with activation energies of ca. 14—16 kcal/mol (Table 3). These isomerizations are also interpretable in terms of two-ring flip mechanisms.

The stereochemically correspondent triarylcarbenium ions show similar behavior, and isomerizations in these systems can also be explained in terms of two-ring flips. [9,33] However, carbenium ions which have only hydrogen atoms in the ortho positions have activation energies of ca. 10—14 kcal/mol [34] (Table 3). It follows that in systems with comparable substituent patterns, i.e., in systems whose ortho ligands offer comparable steric encumbrance, triarylcarbenium ions have substantially higher barriers than the corresponding triarylboranes. No

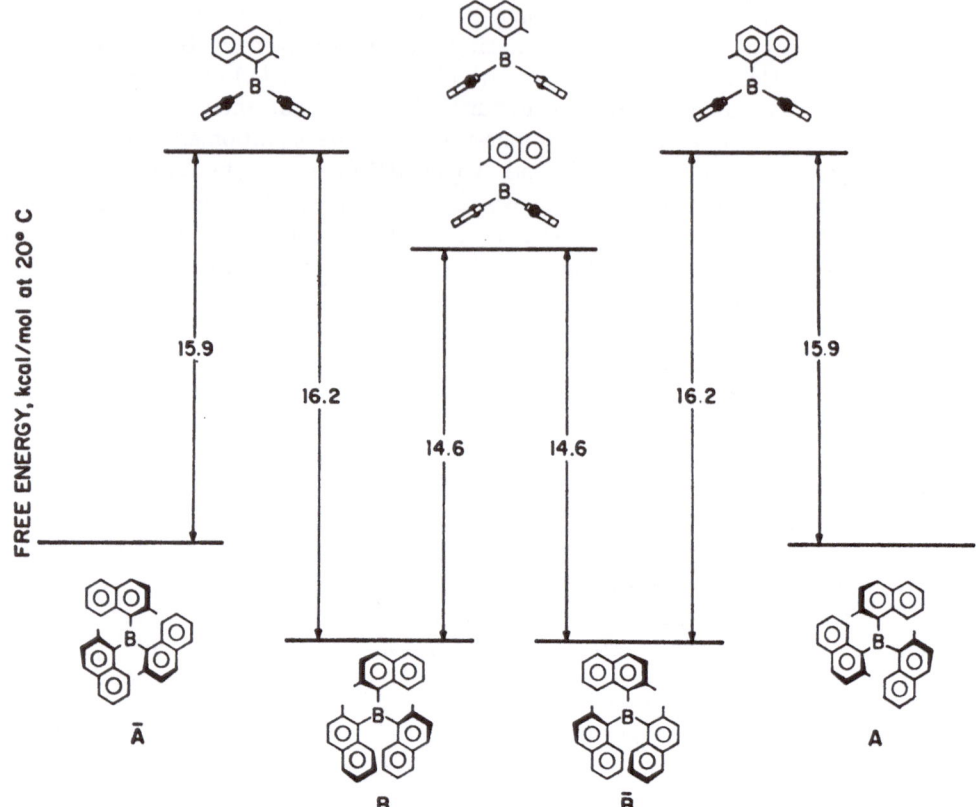

Fig. 10

doubt this is due at least in part to the greater length of the C---B bond (*ca.* 1.58 Å for trimesitylborane [32)]) compared to that of the C---C$^+$ bond (*ca.* 1.45 Å for trityl cation [35)]), a factor which leads to a sterically less crowded structure in the boranes. Differences in conjugative effects are also bound to play some role. It has been reported that in triphenylborane the conjugative interaction between the phenyl rings and boron is negligible (3 $\pm$ 2 kcal/mol). [36)] In addition, in cyclopropyldimethylborane and cyclopropyldifluoroborane the cyclopropyl group is rapidly rotating on the nmr time scale even at −100 °C. [37)] On the other hand, calculations indicate a much higher resonance energy for triarylcarbenium ions. [38)] This comparison suggests that in contrast to the carbenium ions, where conjugative effects seem to play a major role, the barriers to rotation in arylboranes are mainly steric in origin.

Table 3

Triarylcarbenium ions	$\Delta G^{\ddagger}_{25°}$ (kcal/mol)	Triarylboranes	$\Delta G^{\ddagger}$	(T °C)
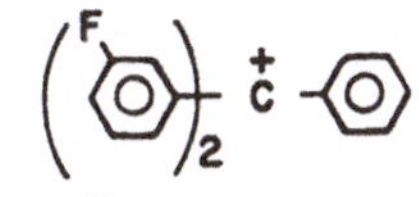	14.10	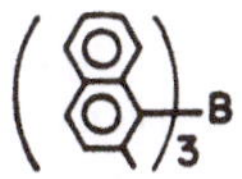	16.2 15.9 14.6	(20) (20) (20)
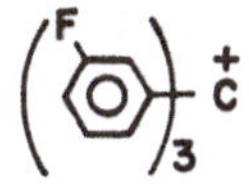	13.10	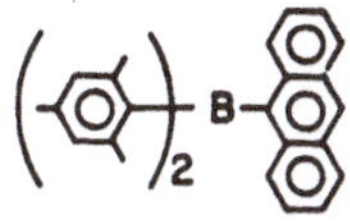	13.9	(15)
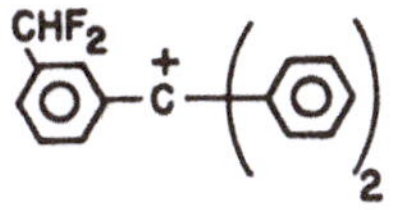	12.7	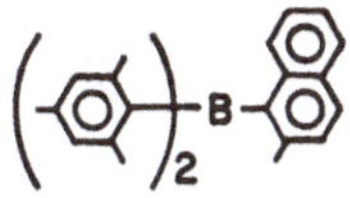	15.4	(20)
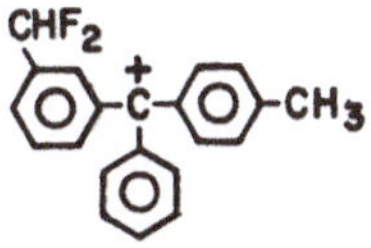	11.6	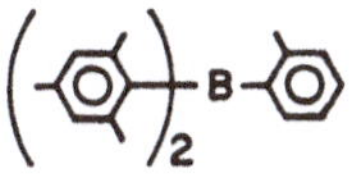	~11.0	(−68)
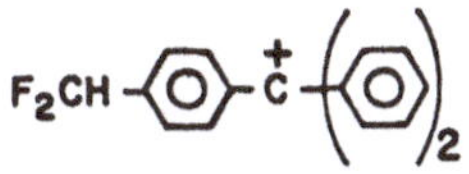	12.3			
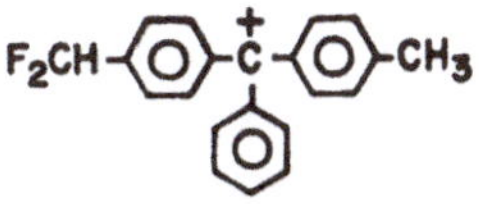	10.4			
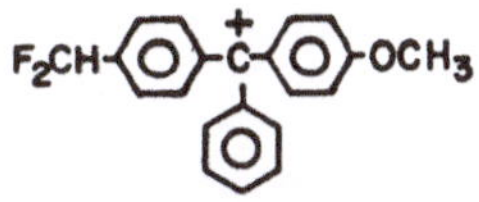	~8.7			

4.2. Trimesitylmethane and Cognates

Trimesitylmethane is an example of isomerism and isomerization in Ar_3ZX systems with C_3 symmetry. Pioneering studies on related triaryl-methanes provided valuable observations concerning structure and mechanism of stereoisomerization. [39–41] In particular, an X-ray diffraction study [41] of dimesityl-(2,4,6-trimethoxyphenyl)methane showed that this compound adopts a propeller conformation in the solid state. A similar conformation was found for triphenylmethane in the gas phase by electron diffraction. [42] Nmr evidence is also consistent with such a conformation for triarylmethanes in solution. [39–41] In the following, we shall briefly describe our experience with trimesitylmethane (7) [43]

23

On the basis of the above-mentioned studies, trimesitylmethane (7) is expected to have a propeller geometry in the ground state, and to exist in two enantiomeric forms, differing only in their sense of twist (helicity). Consequently, each of the three equivalent mesityl rings has two diastereotopic *ortho* methyl substituents, in addition to the *para* methyl group. In the absence of rapid stereoisomerization, the ¹H-nmr spectrum of 7 in an achiral solvent should display three methyl resonances of equal intensity. This expectation is in fact realized at 37 °C,

and the methyl region of the 60-MHz ¹H-nmr spectrum of 7 (in hexachloro-1,3-butadiene solution) is reproduced in Fig. 11.

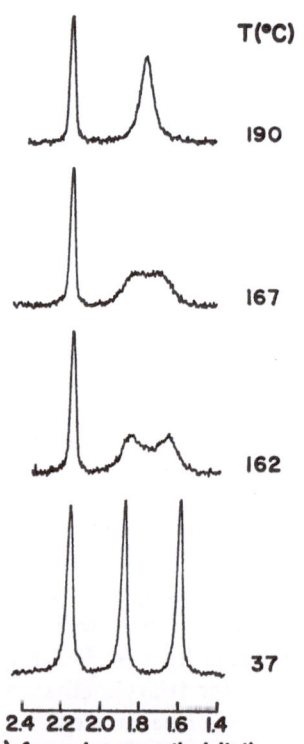

Fig. 11   δ (ppm) from hexamethyldisiloxane

Stereoisomerizations of trimesitylmethane in the propeller conformation may be interpreted in terms of five possible rearrangements: inversion along the C—H bond, and the four flip rearrangements discussed above. The inversion mechanism could involve dissociation into either radicals or ions followed by recombination on the opposite side of the reference plane, or it could involve a concerted displacement reaction. Fig. 12 is a graphical representation of the results of isomerizations of 7 by means of these five rearrangements. The two vertices represent the two enantiomeric propeller conformations, and the edges denote stereoisomerizations. Each of the four flip rearrangements results in enantiomerization, but inversion does not (and hence is strictly a topomerization rather than stereoisomerization). Thus, an optically active sample of 7 would be racemized by any one of the four flip rearrangements, but not by inversion alone.

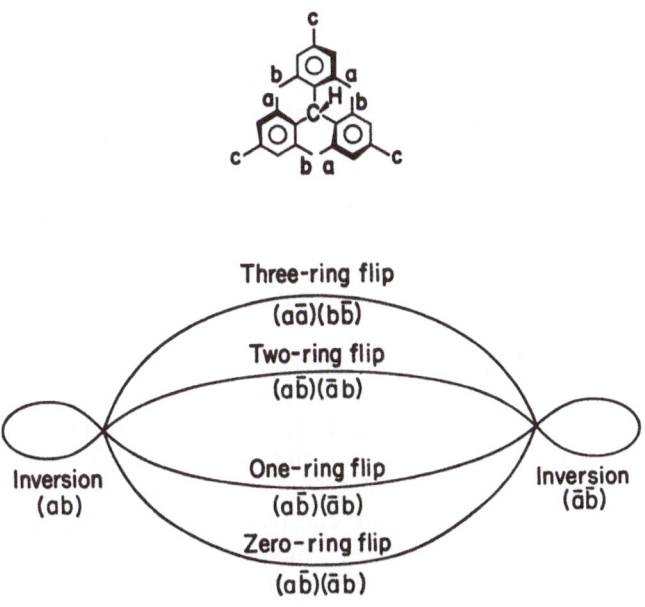

Fig. 12

Each edge shown in Fig. 12 has been labeled with the *ortho* methyl groups whose environments are exchanged by the mechanism in question. Each inversion, zero-, one-, or two-ring flip results in transfer of one or more *ortho* methyl groups from a position proximal to the methine hydrogen to a distal position, and vice versa. The three-ring flip, on the other hand, involves no such transfer, and as a result exchanges only enantiotopic methyl group environments.

25

When the nmr sample of 7 is warmed, the two upfield methyl signals (corresponding to *ortho* methyl groups) broaden and coalesce to a singlet at 167 °C (Fig. 11). This behavior reflects an exchange of environments of the diastereotopic *ortho* methyl groups which becomes rapid on the nmr time scale at 167 °C, and which can be analyzed in terms of the five aforementioned flip rearrangements. Since the three-ring flip does not exchange the proximal and distal methyl groups, it may immediately be ruled out of consideration as an explanation of the nmr spectral behavior, although its concomitant occurrence cannot be rigorously excluded since its operation is undetectable in this experiment. Any one of the remaining four rearrangements is capable of explaining the spectral results.

Since we can gain no further information concerning rearrangement modes from the nmr experiment, we will turn to mechanistic considerations. Inversion is considered unlikely for 7 since it would involve breaking of the central C—H bond. It might be expected that inversion pathways could become important in compounds where this hydrogen is replaced by a readily ionizable group such as halogen or hydroxide, although arguments against this pathway in related compounds have been presented.[40] Furthermore, the zero-ring flip transition state is without a doubt much more sterically crowded than the transition states for the one- or two-ring flips. Accordingly, the zero-ring flip may be eliminated on steric grounds as an interpretation of the spectral results, and we are thus left with the one- and two-ring flips as possible explanations for the observed coalescence of the *ortho* methyl group resonances of 7.

We turn to empirical force field calculations in order to choose between these two mechanisms. [44] Such calculations indicate that the two-ring flip mechanism is the lowest-energy pathway, and yield a barrier of 20 kcal/mol for the two-ring flip of 7. [44] The experimental free energy of activation for stereoisomerization derived from the temperature-dependent ^{1}H-nmr spectrum, is $\Delta G^{\ddagger}_{167}$ 21.9 kcal/mol [43], in excellent agreement with the calculated value. This high barrier admits of the possibility that 7 is separable into its optical antipodes at moderately low temperatures.

Triarylsilanes have also been shown to adopt propeller conformations in the solid state [45] and in solution. [46] Trimesitylsilane (8) exhibits a temperature dependent ^{1}H-nmr spectrum which indicates rapid stereoisomerization at ambient temperature. Thus, at 40 °C the spectrum features two singlets in the methyl region in a ratio of 1:2, assigned to the *para* and *ortho* methyl substituents, respectively. Upon cooling the sample, the *ortho* methyl proton signal broadens and splits into two signals of equal intensity, a result consistent with a propeller conformation on the nmr time scale. Trimesitylsilane obviously is stereochemically correspondent to trimesitylmethane and therefore the analysis given above applies with equal force in this case. Assuming that 8 also stereoisomerizes by the two-ring flip mechanism, the barrier to enantiomerization (derived from the temperature-dependent ^{1}H-nmr results) is $\Delta G^{\pm}_{47}$ 10.9 kcal/mol. [46]

The stereochemical correspondence between trimesitylmethane and trimesitylsilane extends to other Group IVA elements. Thus the barrier to enantiomerization by a two-ring flip mechanism for trimesitylgermane is $\Delta G^{\pm}_{80}$ 9.2 kcal/mol. [47]

The above results bring to light an interesting correlation: the barriers to enantiomerization of the trimesityl-ZH systems decrease as Z progresses down the periodic table in Group IVA. Doubtless this trend is largely due to the increasing Z-aryl bond length. Other factors such as bond angles and bond stretching force constants could also affect the barriers, but are expected to play a minor role.[l]

Acknowledgement: We thank the National Science Foundation for their generous support of this research.

5. References

[1] Webster's Third New International Dictionary of the English Language, Unabridged. Springfield, Mass.: G. and C. Merriam Co. 1966.

[2] Farag, M. S.: Acta Cryst. 7, 117 (1954).

[3] Bastiansen, O.: Acta Chem. Scand. 3, 408 (1949).

[4] Eaton, S. S., Hutchison, J. R., Holm, R. H., Muetterties, E. L.: J. Am. Chem. Soc. 94, 6411 (1972).

[5] Ermer, O., Gerdil, R., Dunitz, J. D.: Helv. Chim. Acta 54, 2476 (1971).

[6] Sundaralingam, M., Jensen, L. H.: J. Am. Chem. Soc. 88, 198 (1966).

[7] Bart, J. C. J.: Acta Cryst. 24B, 1277 (1968).

[8] Newman, A. C. D., Powell, H. M.: J. Chem. Soc. 3747 (1952).

[9] Gust, D., Mislow, K.: J. Am. Chem. Soc. 95, 1535 (1973).

[10] Kurland, R. J., Schuster, I. I., Colter, A. K.: J. Am. Chem. Soc. 87, 2279 (1965).

[11] Ruch, E., Ugi, I.: Theor. Chim. Acta 4, 287 (1966). — Ruch, E., Ugi, I.: Topics Stereochem. 4, 99 (1969).

[12] Muetterties, E. L.: J. Am. Chem. Soc. 90, 5097 (1968); 91, 1636 (1969).

[13] Gielen, M., Depasse-Delit, C.: Theor. Chim. Acta 14, 212 (1969). — Gielen, M., Mayence, G., Topart, J.: J. Organometal. Chem. 18, 1 (1969). — Gielen, M., Topart, J.: J. Organometal. Chem. 18, 7 (1969). — Gielen, M., Vanlautem, N.: Bull. Soc. Chim. Belges 79, 679 (1970).

---

[l] A similar trend has been reported for trimesitylphosphine, arsine, stibine, and bismuthine. [40]

K. Mislow, D. Gust, P. Finocchiaro, and R. J. Boettcher

[14] Eaton, S. S., Eaton, G. R.: J. Am. Chem. Soc. *95*, 1825 (1973).
[15] Klemperer, W. G.: J. Chem. Phys. *56*, 5478 (1972); J. Am. Chem. Soc. *95*, 2105 (1973).
[16] Musher, J. I.: Inorg. Chem. *11*, 2335 (1972).
[17] Ruch, E., Hässelbarth, W.. Theor. Chim. Acta *29*, 259 (1973).
[18] Finocchiaro, P., Gust, D., Mislow, K.: J. Am. Chem. Soc., in press.
[19] Ginsburg, D.: Acc. Chem. Res. *2*, 121 (1969).
[20] Harris, M. M.: Progress in Stereochemistry, p. 174. (eds. W. Klyne and P. B. D. de la Mare), Vol. 2. London: Butterworths 1958. — Ollis, W. D., Sutherland, I.'O.: Chem. Commun., *1966*, 402.
[21] Gust, D., Finocchiaro, P., Mislow, K.: Proc. Natl. Acad. Sci. U.S. *70*, 3445 (1973).
[22] Berry, R. S.: J. Chem. Phys. *32*, 933 (1960).
[23] Musher, J. I.: J. Am. Chem. Soc. *94*, 5662 (1972).
[24] Ugi, I., Marquarding, D., Klusacek, H., Gillespie, P., Ramirez, F.: Acc. Chem. Res. *4*, 288 (1971).
[25] Mislow, K.: Acc. Chem. Res. *3*, 321 (1970).
[26] Hellwinkel, D.: Chimia *22*, 488 (1968), and references cited therein.
[27] Casey, J. P., Mislow, K.: Chem. Commun. *1970*, 1410.
[28] Houalla, D., Wolf, R., Gagnaire, D., Robert, J. B.: Chem. Commun. *1969*, 443.
[29] Hellwinkel, D.: Angew. Chem. Intern. Ed. Engl. *5*, 725 (1966). — Hellwinkel, D.: Chem. Ber. *99*, 3628, 3642, 3660, 3668 (1966).
[30] Whitesides, G. M., Bunting, W. M.: J. Am. Chem. Soc. *89*, 6801 (1967).
[31] Finocchiaro, P.: unpublished results.
[32] Blount, J. F., Finocchiaro, P., Gust, D., Mislow, K.: J. Am. Chem. Soc. *95*, 7019 (1973).
[33] Rakshys, Jr., J. W., McKinley, S. V., Freedman, H. H.: J. Am. Chem. Soc. *93*, 6522 (1971).
[34] Schuster, I. I., Colter, A. K., Kurland, R. J.: J. Am. Chem. Soc. *90*, 4679 (1968). — Rakshys, Jr., J. W., McKinley, S. V., Freedman, H. H.: J. Am. Chem. Soc. *92*, 3518 (1970).
[35] Gomes de Mesquita, A. H., MacGillavry, C. H., Eriks, K.: Acta Cryst. *18*, 437 (1965).
[36] Galuashvili, Zh. S., Romm, I. P., Gur'yanova, E. N., Viktorova, I. M., Sheverdina, N. I., Kocheshkov, K. A.: Doklady Akad. Nauk SSSR *207*, 99 (1972).
[37] Cowley, A. H., Furtsch, T. A.: J. Am. Chem. Soc. *91*, 39 (1969).
[38] Strohbusch, F.: Tetrahedron *28*, 1915 (1972). — Gold, V.: J. Chem. Soc. *1956*, 3944. — Shanshal, M., J. Chem. Soc. Perkin II *1972*, 335.
[39] Kessler, H., Moosmayer, A., Rieker, A.: Tetrahedron *25*, 287 (1969).
[40] Rieker, A., Kessler, H.: Tetrahedron Letters *1969*, 1227.
[41] Sabacky, M. J., Johnson, S. M., Martin, J. C., Paul, I. C.: J. Am. Chem. Soc. *91*, 7542 (1969).
[42] Andersen, P.: Acta Chem. Scand. *19*, 622 (1965).
[43] Finocchiaro, P., Gust, D., Mislow, K.: J. Am. Chem. Soc., in press.
[44] Andose, J. D., Mislow, K.: J. Am. Chem. Soc., in press.
[45] Chieh, P. C., Trotter, J.: J. Chem. Soc. (A), *1969*, 1778.
[46] Boettcher, R. J., Gust, D., Mislow, K.: J. Am. Chem. Soc., *95*, 7157 (1973).
[47] Boettcher, R. J.: unpublished results.

Received August 6, 1973

On the Helicity of Variously Twisted Chains of Atoms

Professor James H. Brewster

R. B. Wetherill Laboratory of Chemistry, Purdue University,
West Lafayette, Indiana, USA

Contents

I. Introduction

It was shown early on by Fresnel [1] that an optically active substance has a slightly different refractive index for right and left circular polarized light and he suggested that this requires a chiral microstructure, possibly helical in nature. It might seem, at first sight, that the notion of helicity in small molecules would be rendered inoperative or unnecessary by the highly productive suggestion of van't Hoff [2] and le Bel [3] that carbon atoms are tetrahedral and will act as centers of structural chirality when they are asymmetrically substituted. It is, indeed, difficult to discern any helicity in a chiral five atom system such as *1*. But the concept of

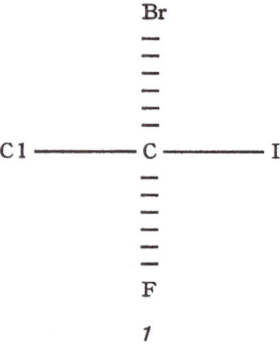

1

helicity has haunted those who have worried about the optical rotation that is such a characteristic property of chiral systems that the *structural* phenomenon of enantiomerism became known as *optical* isomerism. Thus, Gibbs [4], in an offhand comment, and Drude [5], in a detailed theory, suggested that dextrorotation would result if a charged particle were to move in a right-handed helical path under the influence of the electric field of plane polarized light. Such motion would produce a magnetic moment parallel to the electric moment [6]. Levorotation would result when the electric and magnetic moments were antiparallel due to a left-handed helical motion. Kuhn [7] and Kirkwood [8] showed that the electrons of a pair of chromophores could, if coupled chirally, give such parallel or anti-parallel electric and magnetic montents. Kauzmann [9]

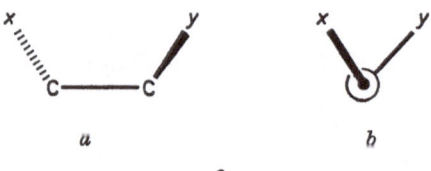

a   *b*

2

presented a model in which twisted bond systems (*e.g.*, as in *2*) act as helical conductors for the motion of electrons. It has been found, indeed, that collections of helices made of copper wire are, in complete macro-analogy, capable of rotating the plane of polarized radio-frequency electromagnetic radiation.[10,11] We[12] have further adapted this model to allow a direct application to organic molecular systems, suggesting the basic equation:

$$[M]_D = 652 \frac{LA}{D^2} (\Sigma R_D) f(n) \qquad (1\,a)$$

where (see Fig. 1):

> $[M]_D$ is the contribution of the particular chain to molecular rotation (Na_D line),
>
> L is the end-to-end distance (Å) of the chain,
>
> A is the area (Å²) subtended on a plane perpendicular to the end-to-end line,
>
> D is the sum of the bond lengths in the chain,
>
> ΣR_D is the sum of bond refractions, and
>
> $f(n)$ is a function of the refractive index of the whole solution.

The use of this model requires a detailed conformational analysis of flexible molecules to determine whether a given twisted chain of atoms shows a net preference for right- or left-helicity under the influence of chiral centers. Thus, R-2,2,3-trimethyl pentane (*3*) should exist pre-dominantly in the conformation shown, with the twisted chain of four carbon atoms forming a right-handed helix. The substance is, in fact, dextrorotatory.[13]

3

It becomes important, thus, to consider in some detail the geometric attributes of twisted chains of atoms regarded as irregular helices, this being one of the significant ways in which the chirality of asymmetric atoms finds structural expression and becomes manifest. The present discussion is largely confined to the geometric aspects of this subject; this leads, however, to the conclusion that the helical conductor model may require some modification before it can be applied to relatively large helices (Section VI). It also leads to the conclusion that the helicity of a long chain, considered as a single unit (the "helical conductor model" [12]), may differ from that obtained by considering it as a collection of smaller helical units (the "conformational dissymmetry model" [14]) — not only in magnitude, but in sign. This lays the basis for possibly productive tests of these two models.

II. The Helicity of Crooked Lines

A helix is the figure generated by the motion of a point *around* and *along* a line — the helix axis. Archetypically, this axis is a straight line and the two kinds of motion are, respectively, *circular* and *linear*, at a constant distance from the axis. The helix is *dextroverse* (right-handed) if the overall motion is *clockwise* and *away* from the observer (or counterclockwise and toward) and *sinistroverse* if *counterclockwise* and *away*. In a right-handed coordinate system such a helix, if dextroverse, obeys the equations:

$$x = r \cos 2\pi \frac{L}{P_t} \tag{2a}$$

$$y = r \sin 2\pi \frac{L}{P_t} \tag{2b}$$

$$z = L \tag{2c}$$

where r is the *radius* of the circular motion, L is the *linear* progress along the helix axis (z) and P_t is the *pitch*, the axial distance corresponding to one complete *turn* (Fig. 1). Where t is the number of *turns*:

$$L = t \cdot P_t. \tag{3}$$

One turn of such a helix subtends a circle on the plane xy and the area (A_t) of this circle is a suitable measure of motion in the x and y directions, as the distance P_t is the measure of motion in the z direction. The product of these two quantities is the volume (V_t) of the cylinder on the surface of which one turn of the helix can be inscribed and thus of the *helix domain* defined by the helix thread:

$$V_t = A_t L \tag{4a}$$

$$= A_t P_t t. \tag{4b}$$

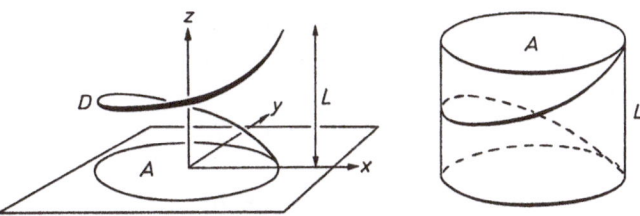

Fig. 1. A single-turn dextroverse helix. L is the end-to-end distance, this line being parallel to the z axis; A is the area subtended on the xy plane; D is the length of the line forming the helix

The first relation will hold for *any* crooked line which is placed so that the end-to-end straight line (L) is parallel to the z axis. If the necessarily closed figure subtended on plane xy is a straight line then the crooked line must lie in a plane and be neither helical nor chiral. If the subtended figure is itself simply a crooked line with no area then the original line encloses no helix domain and the original line is not helical; it may, however, be chiral and it should be stressed that not all chiral lines are helical. When the subtended figure, however complex, consists of a single area that is generated without a crossing of lines then the helix is unequivocally of one or the other handedness. But if there is a crossing of lines (as in Fig. 2 where the cross section is a figure eight) the areas on opposite sides of the crossing-point are the bases of cylinders enclosing domains of opposite helicity and the helix is *amphiverse*. In such cases it may be useful to distinguish the *absolute* helicity — as measured by the total volume of helix domains regardless of chirality — and *net* helicity — as measured by the difference in volume of the right and left helix domains. For our purposes it is the latter which is important.

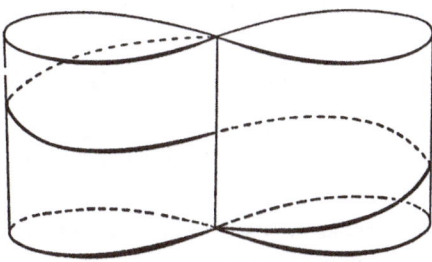

Fig. 2. An amphiverse helix with figure eight cross-section. The two helix domains are of the same size but opposite in sign

For a simple circular helix the thread-length, D, the linear displacement, L, and the circumference of the subtended circle are related as the sides of a right triangle:

$$4 \pi^2 r^2 t^2 = D^2 - L^2 \tag{5}$$

since

$$A_t = \pi r^2 = \frac{D^2 - L^2}{4 \pi t^2} \tag{6}$$

then:

$$V = \frac{L (D^2 - L^2)}{4 \pi t^2}. \tag{7}$$

For any given number of turns:

$$\frac{dV}{dL} = \frac{D^2}{4 \pi t^2} - \frac{3 L^2}{4 \pi t^2} \tag{8}$$

whence maximum volume occurs when

$$L = \frac{D}{\sqrt{3}} \tag{9}$$

or,

$$(V_{max})_t = \frac{D^3}{6\sqrt{3} \, \pi t^2}. \tag{10}$$

For reasons to be discussed below, we limit consideration to cases where $t \geqslant 1$;

$$V_{max} = \frac{D^3}{6\sqrt{3} \, \pi}. \tag{11}$$

It is evident that this volume will be larger where the subtended figure is a circle than in any other case and that this volume corresponds to the largest helix domain that can be enclosed by a line of length D. We may, thus, define an *index of helicity* which can serve as a measure of the extent to which the helix domain enclosed by *any* crooked line of length D (considered as a helix with $t \geqslant 1$) approaches this maximum value:

$$H = \frac{V}{V_{max}} = 6\sqrt{3} \, \pi \, \frac{LA}{D^3}. \tag{12}$$

The index of helicity can be used in an alternative formulation of the basic equation for the helical conductor model of optical activity (1):

$$[M]_D = 20 \cdot H \cdot D \cdot \Sigma(R_D) \cdot f(n). \tag{13}$$

III. Short Chains of Atoms

In our original analysis [14] we considered it likely that the asymmetric atom with four different attachment atoms (as in *1* but not in *3*) and the twisted chain of four atoms (*2*) were the two main structural units contributing to long-wavelength optical rotation. It was recognized that larger systems might also be important and the helical conductor model [12] was developed in an effort to deal with them. In this model a twisted chain of, say, six atoms would be regarded as an indivisible single helix, whereas in the earlier model it would have been treated as three overlapping four atom units. It has been a part of our current concern to determine whether these two ways of considering such a chain are at all equivalent. We find that in some cases they are but in others they are not.

Fig. 3—15 show four-, five- and six-atom chains ($A-C_{n-2}-B$) in their non-planar staggered conformations (dihedral angles of 60° and 180°). The individual bond conformations are denoted: P (positive), M (minus), consistent with the proposals of Cahn, Ingold and Prelog [15], and T (*trans*), as shown in *4*. Using the familiar properties of triangles and the tetrahedral bond angle ($\cos \tau = -1/3$) (see Ref. 12 for the derivation of the equation in Fig. 3) we have derived expressions for the subtended areas (A) as needed for use with the helical conductor model (Eq. (1)). It turns out that all of these expressions contain the term L

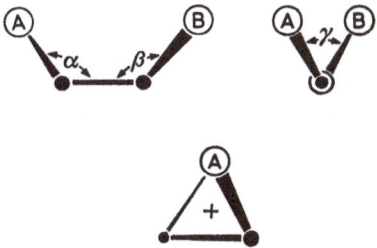

Fig. 3. Four atom chain in the absolute conformation P.

Subtended area:

$$\frac{\sin \alpha \sin \beta \sin \gamma}{2} \frac{d_A\, d_B\, d_C}{L}$$

With tetrahedral bond angles and staggered conformations:

$$\frac{2}{3\sqrt{3}} \frac{d_A\, d_B\, d_C}{L}$$

When $d_A = d_B = d_C$: $H = \dfrac{4\pi}{27} = 0.4654$

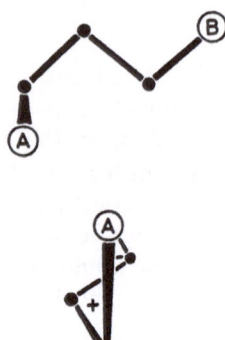

Fig. 4. Five atom chain in the absolute conformation PT.

Subtended areas:

dextroverse: $\dfrac{2\, d_A\, d_C^3}{3\sqrt{3}\,(d_B + d_C)\, L}$

sinistroverse: $\dfrac{2\, d_A\, d_C\, d_B^2}{3\sqrt{3}\,(d_B + d_C)\, L}$

net: $\dfrac{2\, d_A\, d_B\,(d_C - d_B)}{3\sqrt{3}\, L}$

When $d_A = d_B = d_C$: $H = 0$

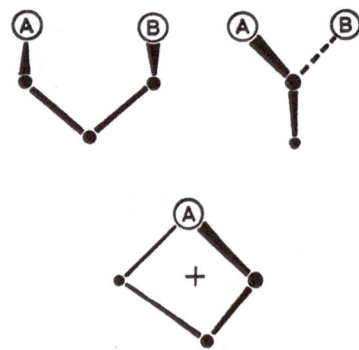

Fig. 5. Five atom chain in the absolute conformation PP.

Subtended area: $\dfrac{2\, d_C\,(d_A\, d_C + 2\, d_A\, d_B + d_B\, d_C)}{3\sqrt{3}} \cdot \dfrac{1}{L}$

When $d_A = d_B = d_C$: $H = \dfrac{\pi}{4} = 0.7854$

Fig. 6. Five atom chain in the absolute conformation PM.

Subtended area: $\dfrac{2\,d_C^2\,(d_A - d_B)}{3\sqrt{3}\ \ L}$

When $d_A = d_B = d_C$: $H = 0$

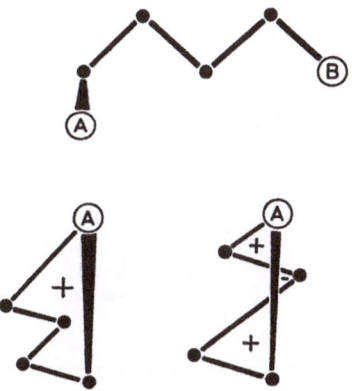

Fig. 7. Six atom chain in the absolute conformation PTT.

Subtended area: $\dfrac{4\,d_A\,d_B\,d_C}{3\sqrt{3}\ \ L}$

When $d_A = d_B = d_C$: $H = \dfrac{8\,\pi}{125} = 0.2011$

37

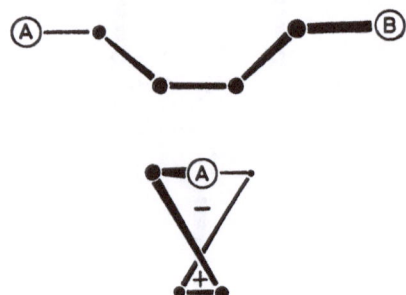

Fig. 8. Six atom chain in the absolute conformation TPT.

Subtended area: $\dfrac{2\,d_C^2}{3\sqrt{3}}\dfrac{[d_C - (d_A + d_B)]}{L}$

When $d_A = d_B = d_C$: $H = -\dfrac{4\,\pi}{125} = -\,0.1005$

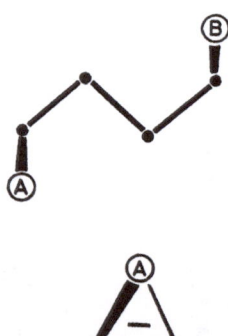

Fig. 9. Six atom chain in the absolute conformation PTP.

Subtended area: $-\dfrac{2\,d_A\,d_B\,d_C}{\sqrt{3}}\dfrac{1}{L}$

When $d_A = d_B = d_C$: $H = -\dfrac{12\,\pi}{125} = -\,0.3016$

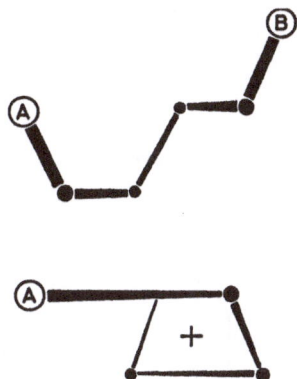

Fig. 10. Six atom chain in the absolute conformation PPT.

Subtended area: $\dfrac{2\,d_C^2\,(3\,d_A + d_C - d_B)}{3\sqrt{3}\qquad L}$

When $d_A = d_B = d_C$: $H = \dfrac{12\,\pi}{125} = 0.3016$

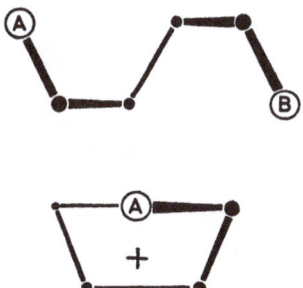

Fig. 11. Six atom chain in the absolute conformation PPP.

Subtended area: $\dfrac{2\,d_C^2\,(3\,d_A + 3\,d_B + d_C)}{3\sqrt{3}\qquad L}$

When $d_A = d_B = d_C$: $H = \dfrac{28\,\pi}{125} = 0.7037$

39

J. H. Brewster

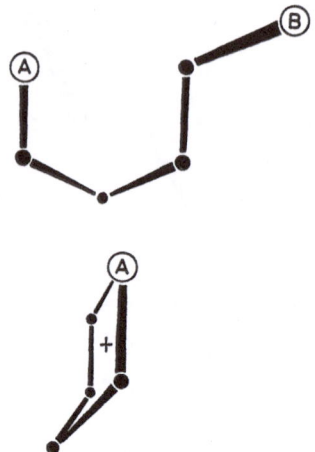

Fig. 12. Six atom chain in the absolute conformation PMT.

Subtended area:
$$\frac{2\ d_C\ [(d_B + d_C)\ (d_A + d_C) - 2\ d_C^2]}{3\sqrt{3}\qquad\qquad L}$$

When $d_A = d_B = d_C$: $H = \dfrac{8\ \pi}{125} = 0.2011$

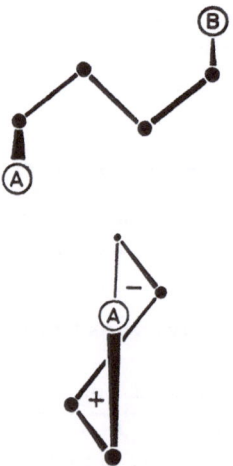

Fig. 13. Six atom chain in the absolute conformation PTM. Subtended area: 0
(all values of d_A and d_B)

40

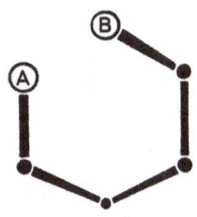

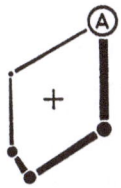

Fig. 14. Six atom chain in the absolute conformation PMP.

Subtended area: $\dfrac{2\ d_C\ [(d_B + d_C)\ (d_A + d_C) - 2\ d_C^2]}{3\sqrt{3}\qquad\qquad L}$

When $d_A = d_B = d_C$: $H = \dfrac{8\pi}{125} = 0.2011$

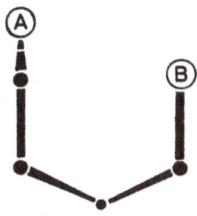

Fig. 15. Six atom chain in the absolute conformation PPM.

Subtended area: $\dfrac{2\ d_C^2\ (3\ d_A + d_C - d_B) + 2\ d_A\ d_B\ d_C}{3\sqrt{3}\qquad\qquad L}$

When $d_A = d_B = d_C$: $H = \dfrac{20\pi}{125} = 0.5027$

J. H. Brewster

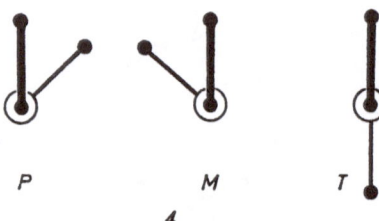

P   M   T

4

in the denominator so that this need not be determined at all. It further turns out that relatively simple expressions can be obtained allowing the terminal bond distances ($C-A=d_A$; $C-B=d_B$) to vary but keeping all internal bond distances ($C-C=d_C$) the same. We have further calculated the index of helicity (H) and the rotatory contribution of the *main chain* for paraffinic chains for each system using Eq. (1a). The large values so obtained have, in earlier applications [12,16], been balanced off by other large values giving, in the end, reasonable final values; it does, however, leave us in the position of calculating rotations as small differences of large numbers — a position that may prove perilous. As will be seen in Section V, there may be reason to use the *average* bond refraction rather than the sum of these values (Eq. (1b), Section V). The values obtained in this way are also shown in Table 1. In many in-

Table 1. Rotatory contributions of relatively short twisted paraffinic chains

Conformation (figure)	Excess of P conformation	Index of helicity (H)	Rotatory contribution as calculated by:		
			Eq. (1a)[1] (Section I)	Eq. (1b)[2] (Section V)	Conformational[3] dissymmetry
P(3)	1	0.4654	747.	+ 69.9	+ 60
PP(5)	2	0.7854	1840.	+157.4	+120
PM(6)	0	0	0	0	0
PTT(7)	1	0.2011	639.	+ 50.3	+ 60
PPP(11)	3	0.7037	2237.	+176.3	+180
PTM(13)	0	0	0	0	0
PMP(14)	1	0.2011	639.	+ 50.3	+ 60
PT(4)	1	0	0	0	+ 60
TPT(8)	1	−0.1005	− 319.	− 25.1	+ 60
PTP(9)	2	−0.3016	− 959.	− 75.6	+120
PPT(10)	2	+0.3016	+ 959.	+ 75.6	+120
PMT(12)	0	+0.2011	+ 639.	+ 50.3	0

[1]) Single helix, *sum* of bond and octet refractions $f(n) = 1.257$ ($n=\sqrt{2}$).
[2]) Single helix, average of bond refractions $f(n) = 1.257$ ($n=\sqrt{2}$).
[3]) P conformation ~60°, M ~−60°, T ~0°.

stances these values closely parallel those obtained by simply letting each skewed unit of three bonds (2) have a rotatory value of $\pm 60°$ for a carbon chain, as in the "conformational dissymmetry model".[14] Note, however, that such use of Eq. (1b) would require neglect of the contribution of hydrogen atoms. The two general approaches are seen, however, to be clearly different for certain other systems, shown at the bottom of the table. Here there are two cases where chains containing only positive bond conformations show negative rotations by the helical conductor model (see also Section IV). This difference should permit decisive tests between the two approaches. We have presented preliminary evidence, for the case of twistane [16], indicating that Eq. (1a) gives better results than the conformational dissymmetry model in systems with C_2 axes. (Reasonably good results are also obtained for that system by confining attention only to those helices having terminal atoms and using only bond refractions. Calculations using Eq. (1b) give poor results.)

IV. Many-turn Helices — Long Chains of Atoms

1. General Considerations

In the preceding section we showed that a relatively short twisted chain of atoms can be treated as a one-turn (or, in some cases, two-turn) helix with an *arbitrary* axis parallel to the end-to-end line. In this section we turn attention to helices having many turns in which an *actual* axis can be discerned. It is necessary at this point to make some distinctions among several different types of helix that may be of significance in organic chemistry.

Line helices (*chordal helices*) are composed of dimensionless points. A chordal helix is *uniform* if all points are equivalent; such a helix will have a circular cross-section and could be inscribed on a right circular cylinder. A nonuniform chordal helix is *regular* if it contains *motifs* repeated in a definite pattern and *irregular* if the motifs are not so repeated. For our purposes, chordal helices are useful mathematical abstractions.

Chain helices (*catenal helices*) are composed of beaded lines in which particular points, separated by line segments, are distinguished in function from others. Here we may distinguish *iterative* chains, in which the motifs of points and lines are repeated regularly, and *non-iterative* chains in which no such repetition occurs. The *backbones* of linear homopolymers are *iterative chains* which may be imagined to assume, on the one hand, two kinds of regular conformation — the zig-zag chain (all dihedral angles 180°) and the set of regular helices — and, on the other, a set of irregular conformations commonly referred to as the *random coil*. We are concerned here with the regular conformations of such helices.

Ribbed helices (*costal helices*) are important in organic chemistry because linear polymers contain *side chains* as well as backbones. We may, then, discern not only the *catenal helix* of the backbone, but the *intercostal helix* formed by all of the ribs and the *intracostal* helices of the individual side chains. The intercostal helix may be iterative (as in an isotactic head-to-tail vinyl polymer or homogeneous poly-α-amino acid) or non-iterative (as in a random copolymer, an atactic polymer or typical protein). The intracostal helices can best be analyzed as short-chain crooked lines, as in Section III. Important as costal helicity is, it is secondary to catenal helicity and we therefore limit our attention to the primary helicity, that of long chains. Indeed, we limit our attention to *catenal* helices having chain motifs of *two atoms and two bonds* as found in head-to-tail vinyl homopolymers:

$$
\begin{array}{cc}
W & Y \\
| & | \\
(C-C-)_n \\
| & | \\
X & Z
\end{array}
$$

or polyoxymethylene $(CH_2-O-)_n$. We designate them [2]-helices. In the most general terms, [2]-helices consist of points (atoms) A and B in regular alternation and line segments (bonds) a and b also in regular alternation:

$$A \xrightarrow{a} B \xrightarrow{b} A \xrightarrow{a} B \xrightarrow{b}.$$

If all of the bonds are single, then the principle distinction between them will be the value of the dihedral or conformation angles γ_a and γ_b.

2. Characteristics of the Motifs in [2]-Helices

A set of two motifs, conformationally adjusted to lie on one plane $(\gamma_a = \gamma_b = 0)$ is shown in Fig. 16a. The corresponding motifs seen in projection on the xy plane (z axis = helix axis) are shown in Fig. 16b.

The *bond angles* are designated θ_A and θ_B (Fig. 16a) and, as seen in projection, ϕ_A and ϕ_B (Fig. 16b).

The bond distances are designated d_a and d_b (Fig. 16a) and, as seen in projection, d_a^* and d_b^* (Fig. 16b).

For the case where all bond distances are the same (d), the familiar properties of triangles give the dimensions shown in Fig. 17a and b.

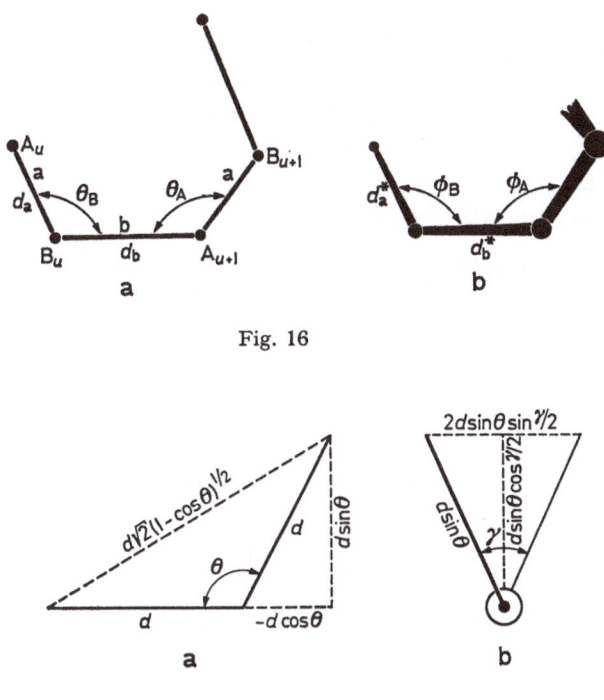

Fig. 16

Fig. 17

3. Characteristics of the Cross Section of [2]-Helices

A segment of the helix cross section as projected on the xy plane is shown in Fig. 18.

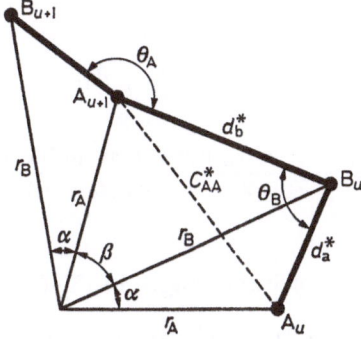

Fig. 18. Cross section of a [2]-helix where the bond angles θ_a and θ_b are different

Letting u be an integer denoting the motif number:

$r_{Au} = r_A$ (radial distance from the helix axis)
$Z_{Au} = u\,(\alpha + \beta)$ (in degrees, read counterclockwise)
$Z_{Au} = u\,(P_a + P_b)$ (P_a and P_b, rise of bonds a and b)
$r_{Bu} = r_B$
$Z_{Bu} = \alpha + u\,(\alpha + \beta)$
$Z_{Bu} = P_a + u\,(P_a + P_b)$.

Each bond subtends an internal triangle, the areas of which are:

$$A_a = \frac{r_A\,r_B}{2}\sin \alpha \qquad (14\,a)$$

$$A_b = \frac{r_A\,r_B}{2}\sin \beta \qquad (14\,b)$$

whence the total area of one cross section, subtended by one turn:

$$A_t = \frac{u}{t}\,\frac{r_A\,r_B}{2}\,(\sin \alpha + \sin \beta) \qquad (14\,c)$$

where u/t is the number of motifs (u) per turn (t). This term is often expressed as a ratio of integers:

$$\frac{u}{t} = \frac{u_C}{t_C} \qquad (15)$$

where u_c is the number of motifs per crystallographic repeat unit and t_c the number of turns per crystallographic repeat unit.

It is evident that:

$$L_t = \frac{u}{t}\,(P_a + P_b) \qquad (16)$$

and:

$$D_t = \frac{u}{t}\,(d_a + d_b) \qquad (17)$$

whence:

$$H = 6\sqrt{3}\,\pi\,\frac{r_A\,r_B}{2}\,\frac{(\sin \alpha + \sin \beta)\,(P_a + P_b)}{u/t\,(d_a + d_b)^3}\,. \qquad (18)$$

4. Effect of Differences of Bond Angle in [2]-Helices

The projected chord C^*_{AA} (Fig. 18) can be defined in three ways:

$$C^{*2}_{AA} = 2\,r^2_A\,[1 - \cos(\alpha + \beta)] \qquad (19\,a)$$

$$C_{AA}^{*2} = d_a^2 + d_b^2 - 2\,d_a\,d_b \cos \theta_B - P_{ab}^2 \qquad (19b)$$

$$C_{AA}^{*2} = d_a^{*2} + d_b^{*2} - 2\,d_a^*\,d_b^* \cos \phi_B . \qquad (19c)$$

From the first two equations:

$$r_A^2 = \frac{d_a^2 + d_b^2 - 2\,d_a\,d_b \cos \theta_B - P_{ab}^2}{2\,[1 - \cos(\alpha + \beta)]} \qquad (20a)$$

and, similarly:

$$r_B^2 = \frac{d_a^2 + d_b^2 - 2\,d_a\,d_b \cos \theta_A - P_{ab}^2}{2\,[1 - \cos(\alpha + \beta)]} \qquad (20b)$$

whence:

$$r_A^2 - r_B^2 = \frac{d_a\,d_b\,(\cos \theta_A - \cos \theta_B)}{1 - \cos(\alpha + \beta)} \qquad (21)$$

when $\theta_A = \theta_B$, $r_A = r_B$

when $\theta_A \neq \theta_B$, $r_A \neq r_B$.

We focus attention here on systems where both bond angles are the same. In this case all the atoms lie on the surface of a single circular cylinder. (When $r_A \neq r_B$ they would lie on the surface of concentric cylinders, see Fig. 19).

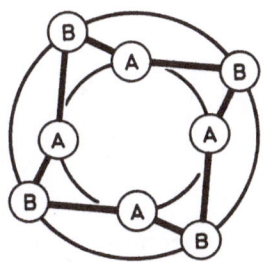

Fig. 19. Cross section of a [2]-helice with $u/t = 4/1$ and having different bond angles, θ_a and θ_b

Since:

$$C_{BB}^{*2} = 2\,r_B^2\,[1 - \cos(\alpha + \beta)] \qquad (19d)$$

it follows that when $r_A = r_B$, $C_{AA}^* = C_{BB}^*$,
and, since:

$$C_{BB}^{*2} = d_a^{*2} + d_b^{*2} - 2\,d_a\,d_b \cos \phi_A \qquad (19e)$$

it follows that when $C_{AA}^* = C_{BB}^*$, $\phi_A = \phi_B$

J. H. Brewster

Thus, for the case where the bond angles are the same, there is but a single value for r, C^* and ϕ. If, in addition, all the bonds are of the same order and thus of the same length (since A—B is the same as B—A), then atoms A and B become equivalent *in the backbone* with bonds a and b differing only in conformation angle. With this simplification each bond acquires local two-fold symmetry with the C_2 axis perpendicular to the helix axis (Fig. 20).

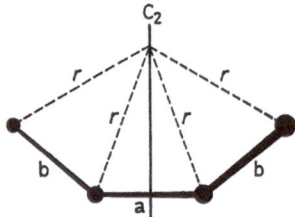

Fig. 20. Cross section of [2]-helix in which $\theta_a = \theta_b$, showing C_2 symmetry about individual bonds

5. Characteristics of the [2]-Helix

A segment of the cross section of a helix ABAB with $\theta_A = \theta_B$ and $d_a = d_b$ is shown in Fig. 21.

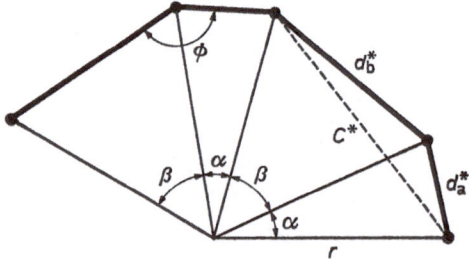

Fig. 21. A segment of the cross section of a [2]-helix with all bond angles the same. All values for the helix radius (r), the motif chord (C) and the external angle (ϕ) are the same. Values for the internal angles (α and β) and the projection bond distances (d_a^* and d_b^*) depend on the conformation angles (γ_a and γ_b)

As shown above all external angles (ϕ) are the same, as are the *chords* (C^*) between atoms, and the radial distance (r) of atoms from the helix axis. Even though the bond distances (d) are the same, those for the projected figure will be different if the dihedral angles around bonds a and b (γ_a and γ_b) are different. We wish now to determine values for

48

r, the radius of the helix cylinder

α and β, the internal angles

ϕ, the external angle

d_a^* and d_b^*, the projection bond distances

P_a and P_b, the rise per bond

C^*, the projection chord

u/t, the helix symmetry or units per turn

and, thence, the helicity index, H, for such a helix in terms of

d, the standard bond length

θ, the standard bond angle

and γ_a and γ_b, the conformation angles.

The following basic relationships obtain:

$$d_a^{*2} + P_a^2 = d^2 \tag{22a}$$

$$d_b^{*2} + P_a^2 = d^2 \tag{22b}$$

$$C^{*2} + (P_a + P_b)^2 = 2\,d^2\,(1 - \cos\theta). \tag{23}$$

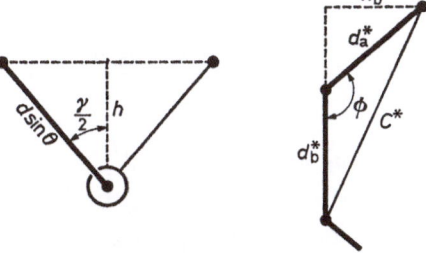

Fig. 22. Relationship between the bond angle (θ) and the external angle of the helix cross section (ϕ)

$$h_b = d \sin\theta \cos(\gamma_b/2) \tag{24a}$$

$$h_b = d_a^* \sin\phi \tag{24b}$$

whence:

$$d_a^* \sin\phi = d \sin\theta \cdot \cos(\gamma_b/2) \tag{25a}$$

and

$$d_b^* \sin \phi = d \sin \theta \cdot \cos (\gamma_a/2) \tag{25b}$$

$$C^{*2} = d_a^{*2} + d_b^{*2} - 2 d_a^* d_b^* \cos \phi . \tag{26}$$

The value for the external angle (ϕ) is the key to all others. The six equations above are in six unknowns and can be solved for $\cos \phi$ as follows:

Combining Eqs. (22a), (22b) and (23) to eliminate P_a and P_b we obtain:

$$2 d^2 \cos \theta = d_a^{*2} + d_b^{*2} - C^{*2} - 2 [d^4 - d^2 (d_a^{*2} + d_b^{*2}) + d_a^{*2} d_b^{*2}]^{1/2} \tag{27}$$

and eliminating C^* by use of Eq. (26):

$$d^2 \cos \theta = d_a^* d_b^* \cos \phi - [d^4 - d^2 (d_a^{*2} + d_b^{*2}) + d_a^{*2} d_b^{*2}]^{1/2} . \tag{28}$$

Substituting Eqs. (25a) and (25b) to eliminate d_a^* and d_b^* and simplifying, we obtain the basic helix equation:

$$\cos \phi = \cos \theta \cdot \cos (\gamma_a/2) \cos (\gamma_b/2) + \sin (\gamma_a/2) \sin (\gamma_b/2) \tag{29}$$

from which the following solutions are obtained:

$$d_a^* = \frac{d \sin \theta \cdot \cos (\gamma_b/2)}{\sin \phi} \tag{30a}$$

$$d_b^* = \frac{d \sin \theta \cdot \cos (\gamma_a/2)}{\sin \phi} \tag{30b}$$

$$P_a = (d^2 - d_b^{*2})^{1/2} \tag{31a}$$

$$P_b = (d^2 - d_b^{*2})^{1/2} \tag{31b}$$

$$C^{*2} = 2 d^2 (1 - \cos \theta) - (P_a + P_b)^2 . \tag{32}$$

Noting that all of the triangles in Fig. 21 are isosceles, it follows that: $\alpha + \beta + 2 \phi = 360°$

$$\text{or} \qquad \frac{\alpha + \beta}{2} = 180° - \phi \tag{33}$$

but

$$r = \frac{C^*/2}{\sin\left(\frac{\alpha+\beta}{2}\right)} = \frac{C^*}{2\sin\phi} \, . \tag{34}$$

The helix symmetry:

$$u/t = \frac{360°}{\alpha+\beta} = \frac{180°}{180-\phi} \, . \tag{35}$$

It is seen from Fig. 21 that:

$$\sin\frac{\alpha}{2} = \frac{d_a^*}{2r} \tag{36a}$$

whence:

$$\sin\alpha = \frac{d_a^*}{2r^2}(4r^2 - d_a^{*2})^{1/2} \, . \tag{36b}$$

Similarly:

$$\sin\beta = \frac{d_b^*}{2r^2}(4r^2 - d_b^{*2})^{1/2} \, . \tag{36c}$$

But, the area subtended by bond a is:

$$A_a = \frac{r^2}{2}\sin\alpha \tag{37a}$$

or

$$A_a = \frac{d_a^*}{4}(4r^2 - d_a^{*2})^{1/2} \tag{37b}$$

or

$$A_a = \pm\frac{d_a^*}{4\sin\phi}[d_b^* - d_a^*\cos\phi] \tag{37c}$$

and, similarly,

$$A_b = \pm\frac{d_b^*}{4\sin\phi}[d_a^* - d_b^*\cos\phi] \tag{37d}$$

It is evident that these areas are to be added when the helix has the simple cross section shown in Fig. 21 whence the motif area:

$$A_m = A_{ab} = \frac{[2\,d_a^*\,d_b^* - (d_a^{*2} + d_b^{*2})\cos\phi]}{4\sin\phi} \, . \tag{38}$$

51

These equations apply directly to what we term *ring* helices (below). The equations appropriate for other kinds of helix are discussed individually below after the next section.

6. General Attributes of [2]-Helices

Conformational maps in which various attributes and properties of [2]-helices are plotted as functions of the two conformation angles, γ_a and γ_b, are shown in Figs. 23—26. The first of these displays the regions, lines and points where qualitatively distinct kinds of helix are found. These helices are shown in cross section in *5—13*.

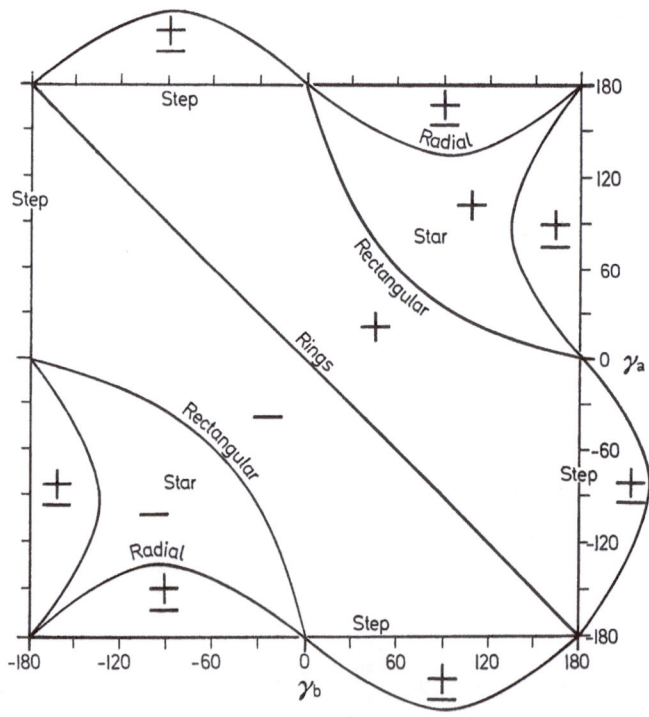

Fig. 23. Conformation map showing points, lines and regions occupied by the various classes of [2]-helix as a function of the two conformation angles

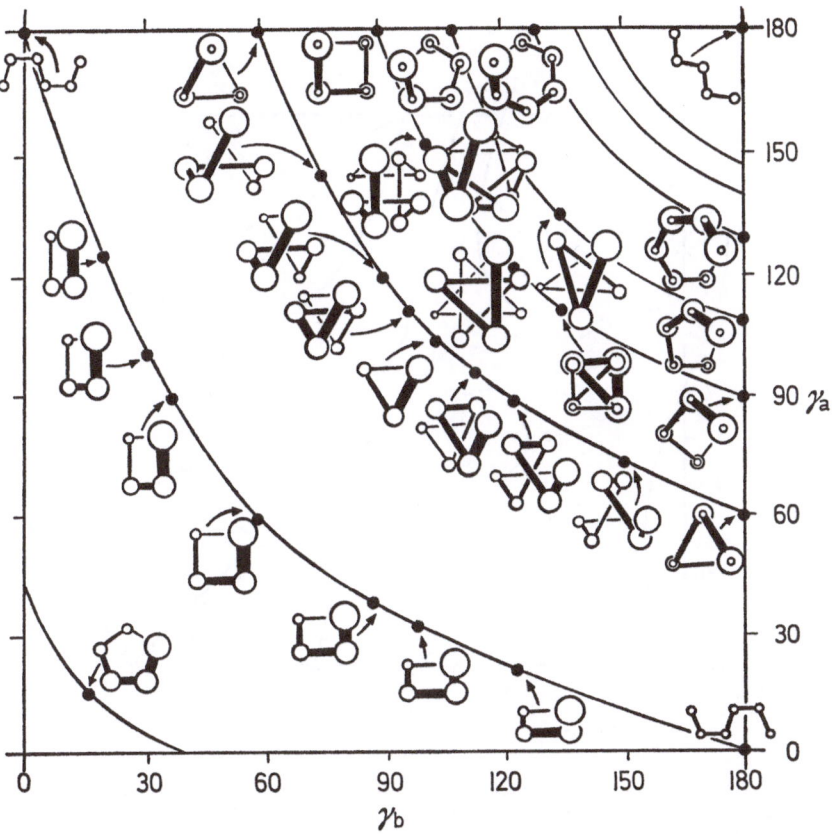

Fig. 24. Conformation map showing detailed cross sections of [2]-helices for cases where both conformation angles are positive

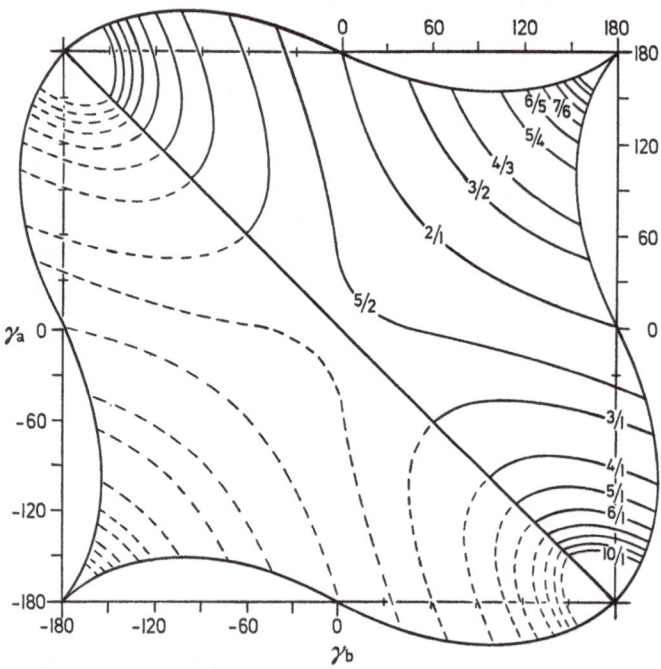

Fig. 25. Conformation map showing contour lines for equal values of helix symmetry (u/t) for [2]-helices as a function of the conformation angles. Positive, or dextroverse, helicity is shown by solid lines (———), negative by broken lines (— — —)

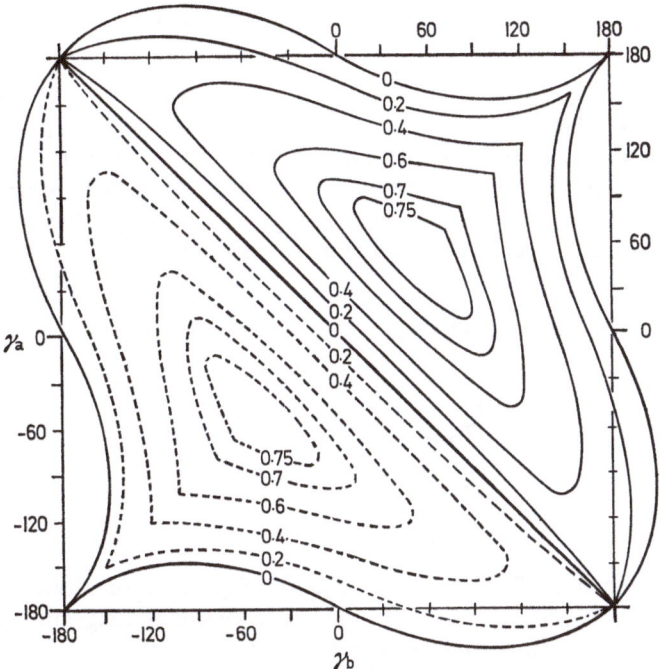

Fig. 26. Conformation map showing contour lines for equal index of helicity (H) for [2]-helices as a function of conformation angle. Solid lines (———) denote positive helicity, broken lines (— — —) denote negative helicity

Several *points of non-helicity* occur, corresponding to the combinations: γ, $\gamma = 0°$, $0°$; γ, $\gamma = 0°$, $180°$; γ, $\gamma = 180°$, $180°$. In real systems of atoms and bonds the first point is forbidden by steric bulk effects and the second (*5*) is disfavored by the strain associated with a *cis* conformation ($\gamma = 0°$). The last point is the conformationally favored all-*trans* planar zig-zag chain (*6*) with symmetry 1/1 as seen end on.

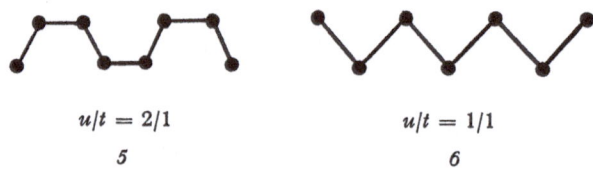

$u/t = 2/1$	$u/t = 1/1$
5	6

There is also a *line of non-helicity* for the case where $\gamma_a = -\gamma_b$. In real systems the points along this line where u/t is an integer are occupied by even-membered rings in their most regular puckered conformations (*e.g.*, *7*). This line divides the largest region of helicity, that of the *ring* helices (*8*), into dextroverse and sinistroverse domains.

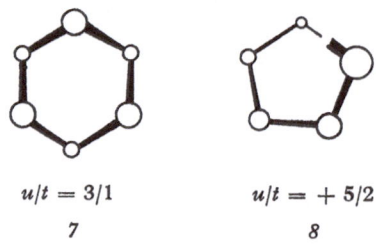

$u/t = 3/1$	$u/t = +5/2$
7	8

A *ring* helix, (*8*), is characterized by *divergence* of alternate bonds (a_1, a_2 and b_1, b_2) as seen in the cross-sectional diagram. Since, in [2]-helices, $\phi_A = \phi_B$ it follows that for *ring* helices:

$$\phi > 90° > \frac{\alpha + \beta}{2} \tag{39a}$$

whence the helix symmetry number is greater than 2/1. The alternate bonds of *star* helices (*9*) *converge* to *include* the helix axis so that:

$$\phi < 90° < \frac{\alpha + \beta}{2} \tag{39b}$$

whence the symmetry number is less than 2/1. An amphiverse helix (*10*) also has *converging* alternate bonds, but they *exclude* the helix axis.

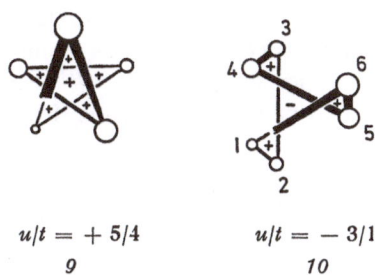

$$u/t = + 5/4$$
9

$$u/t = - 3/1$$
10

The helix symmetry depends on whether handedness is assigned by *motif* ($u/t > 2/1$) or point-by-point ($u/t < 2/1$).

The boundary between star and amphiverse helices is formed by the *radial* helices (*11*), in which one set of alternate bonds converge *at* the helix axis. The *rectangular* helices (*12*), where $\phi = 90°$ and $u/t = 2/1$, separate the star and ring helices and represent one class of helix where alternate bonds are parallel. The *step* helices (*13*) (γ_a or $\gamma_b = 180°$) separate the ring and amphiverse helices and have one set of parallel alternate bonds. Many of these figures are also shown on the conformational map in Fig. 24 for the quadrant where the two conformation angles have the same sign.

Each kind of helix is considered separately below.

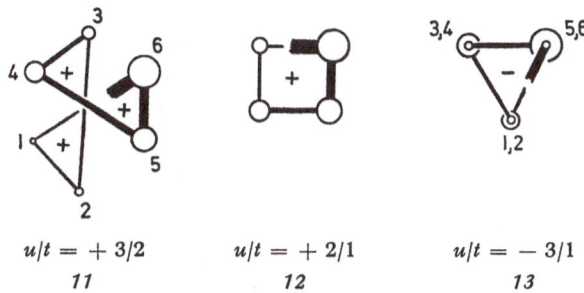

$$u/t = + 3/2$$
11

$$u/t = + 2/1$$
12

$$u/t = - 3/1$$
13

7. Ring Helices and Rings

As seen on the conformational map in Fig. 23 the class of *ring* helices is the largest of all. It completely occupies the two quadrants where the conformation angles are of opposite sign and a significant portion of the other two quadrants as well. In general:

$$\tan (\gamma_a/2) \cdot \tan (\gamma_b/2) < 1/3 . \tag{40}$$

It will be recalled from the previous section that $\phi > 90°$ and $u/t > 2/1$.

The zone of ring helices is divided into two domains of opposite helicity by the line on which closed rings occur. Here $\gamma_a = -\gamma_b$ and only those points are significant where u/t is an integer. The basic equations apply but can be simplified. Thus, from Eq. (29):

$$\cos \gamma = 2 + 3 \cos \phi \tag{41}$$

but,

$$\phi = 180° - \frac{180°}{u/t} \tag{42}$$

so that

$$\cos \gamma = 2 - 3 \cos \left(\frac{180°}{u/t}\right). \tag{43}$$

Thus the conformation angles found in regular puckered rings with even numbers of members can be calculated with ease, see Table 2.

Table 2. Conformation angles of even-membered rings

Members	u/t	$\gamma_a\ (= -\gamma_b)$
6	3	60.00°
8	4	96.97
10	5	115.28
12	6	126.73
14	7	134.66
16	8	140.50
18	9	144.99
20	10	148.56
30	15	159.14
50	25	167.51
100	50	173.76

The *cis* helices, where γ_a or $\gamma_b = 0°$, represent another significant subset of *ring* helices. At one extreme ($\gamma_a = \gamma_b = 0°$) is a figure that is nearly a closed pentagon (14) and at the other is the plane chain with $\gamma = 0°$, $\gamma = 180°$ (5). Here:

$$\cos \phi = -1/3 \cos (\gamma/2) \tag{44}$$

whence:

$$\cos \gamma = 9 \cos \left(\frac{360°}{u/t}\right) + 8 \tag{45}$$

$$L_m = \frac{4}{3} \frac{\sin\ (\gamma/2)}{\sin\ \phi} \qquad (46)$$

$$A_m = -\frac{2\cos\ \phi}{\sin\ \phi} \left[\frac{15 + \cos\ \gamma}{17 - \cos\ \gamma}\right]. \qquad (47)$$

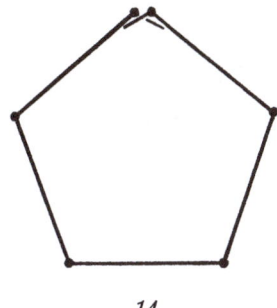

14

Values calculated for γ and H for particular values of u/t are shown in Table 3. These values were used by interpolation to calculate the conformation angles corresponding to particular values of H, as shown in Table 4.

Table 3. *cis*-ring helices ($\gamma_a = 0°$)

u/t	γ_b	H
2.55215	0°	0
2.50	44.04	0.5383
2.45	62.79	0.6693
2.40	78.13	0.7177
2.35	91.87	0.7161
2.30	104.77	0.6798
2.25	117.21	0.6158
2.20	129.45	0.5291
2.15	141.69	0.4231
2.10	154.09	0.2993
2.05	166.80	0.1582
2.00	180	0

8. Star Helices

This class of helices contains the set where $\gamma_a = \gamma_b$ in which the cross section figure contains only lines of one length and the helix symmetry is x/x-1. Here:

$$\cos\ \phi = \frac{1 - 2\cos\ \gamma}{3}. \qquad (48)$$

Table 4. Helicity contour points for *cis*-ring helices ($\gamma_a = 0°$)

H	γ_b	
0	0°	180°
0.10	7.29	171.60
0.20	14.56	163.02
0.30	22.52	153.82
0.40	30.54	143.82
0.50	40.46	132.82
0.60	52.40	119.60
0.70	71.82	97.08

Since:

$$u/t = \frac{180°}{180° - \phi}$$

$$\phi = 180° \frac{u - t}{u}$$

or

$$\cos \phi = \frac{180°}{u}. \tag{49}$$

It is readily shown that:

$$d^* = \frac{\sqrt{2}}{\sqrt{3}} \frac{d}{(1 + \cos \phi)^{1/2}} = \frac{d}{(2 - \cos \gamma)^{1/2}} \tag{50}$$

$$A_m = \frac{(1 - \cos \phi) d^2}{3 \sin \phi (1 + \cos \phi)} \tag{51}$$

$$L_m = 2 \left[\frac{1/3 + \cos \phi}{1 + \cos \phi} \right]^{1/2} d. \tag{52}$$

Values have been calculated for γ and H corresponding to particular values for u/t and are shown in Table 5. From these were calculated values for γ corresponding to particular values for H (Table 6). The areas of the cross section, taken as the sum of the triangles subtended by the lines d^* accurately takes into account the number of turns to a crystallographic repeat.

9. Rectangular Helices

These helices, where $\phi = 90°$, have alternate bonds parallel and divide the *ring* helices, where such bonds diverge ($\phi > 90°$), and the *star* helices,

Table 5. Star helices

u/t	γ	H
2	60°	0.7854
5/3	87.91	0.6347
4/3	124.10	0.3871
5/4	135.52	0.3107
6/5	143.04	0.2610
7/6	148.37	0.2256
8/7	152.35	0.1990
9/8	155.44	0.1781
10/9	157.91	0.1613
25/24	171.18	0.0673
50/49	175.59	0.0356
100/99	177.80	0.0173
1.0	180	0

Table 6. Helicity contour points for ring and star helices ($\gamma_a = \gamma_b$)

H	γ	
	Ring	Star
0	0°	180°
0.10	3.80	166.60
0.20	7.60	152.15
0.30	11.40	137.13
0.40	15.29	122.10
0.50	20.76	107.45
0.60	26.22	92.98
0.70	33.93	77.62
0.75	40.28	67.74
0.7854	60	(Rectangular helix)

where they converge ($\phi > 90°$). The helix symmetry is 2/1 and the relationship of the two conformation angles is:

$$\tan(\gamma_a/2) \cdot \tan(\gamma_b/2) = \tfrac{1}{3}. \tag{53}$$

The line for these helices forms a major contour line in the conformational maps in Figs. 22, 23 and 24.

10. Amphiverse Helices

In this series *amphiverse* helices occur when both of the conformation angles are of the same sign and when one of them approaches 180°. The cross sections are complex, as seen, for example, in *10* and *16*, but

Table 7. Conformational angles for rectangular helices

γ_a	γ_b	γ_a	γ_b
0°	180°	90°	36.87°
15	136.90	105	28.69
30	102.42	120	21.79
45	77.65	135	15.72
60	60	150	10.21
75	46.96	165	5.02

they derive in an obvious manner from those of the step helices (*13, 15*) and of the radial helices (*11, 17*) and these helices form boundary lines for the regions of amphiversity ($\pm$) on the conformational map (Fig. 23).

It will be noted that the central segment of the cross section of an amphiverse helix has a negative helicity (as in *16*) when the conformation angles are positive; this condition obtains in the corresponding *step* helix (*15*). The triangular segments, on the other hand, have positive helicity when the conformation angles are positive, as with the corresponding *radial* helix (*17*). It is evident that at some point the *net* helicity (taking sign into account) must become zero even though the *absolute* helicity (neglecting sign) is appreciable. We take this to be the point at which the *net area* of the cross section of a full crystallographic repeat unit is zero.

The helix symmetry and the handedness of amphiverse helices can be assigned in two ways. In the case shown in *16*, the assignment based on *motifs* (1,2—3,4—5,6—7,8) is − 4/1, as for the related helix *15*. The assignment based on *points* (1—2—3—4—5—6—7—8), however, is +4/3, as for the related radial helix (*17*). Note that this ambiguity does not

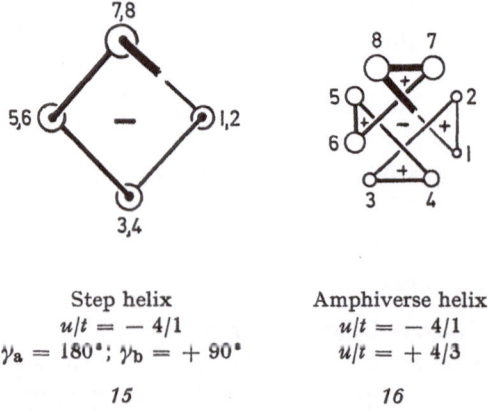

Step helix
$u/t = - 4/1$
$\gamma_a = 180°; \gamma_b = + 90°$

15

Amphiverse helix
$u/t = - 4/1$
$u/t = + 4/3$

16

occur in the step helix but does remain in the radial (*17*) and star helices (*18*) where the *motif* angle is greater than 180°. In these cases assignment of symmetry by motif ($u/t = -4/1$) seems highly artificial, given that it is the *chain of bonds* that is the line forming the helix.

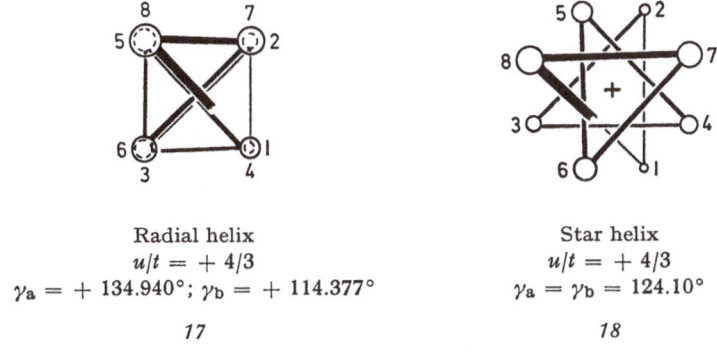

Radial helix	Star helix
$u/t = +4/3$	$u/t = +4/3$
$\gamma_a = +134.940°; \gamma_b = +114.377°$	$\gamma_a = \gamma_b = 124.10°$
17	*18*

As seen in Fig. 27, each motif contains a triangular area subtended by bond a (as Δ_{34y}) with positive helicity and another triangular area

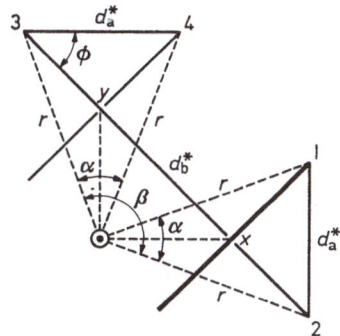

Fig. 27. Segment of the cross section of an amphiverse helix

subtended by a *portion* of bond b (as Δ_{xy0}) with negative helicity. The *net* sum of these areas is A_m and may be positive, negative or zero. It is readily seen that the principles derived above apply here. Thus, the total area subtended by bond a is:

$$A_a = \Delta_{340} = \Delta_{34y} + \Delta_{3y0} + \Delta_{4y0} \tag{54a}$$

and that subtended by *all* of bond b:

$$A_b = \Delta_{120} = \Delta_{xy0} + \Delta_{3y0} + \Delta_{2x0} \tag{54b}$$

but:

$$\Delta_{3y0} = \Delta_{4y0} = \Delta_{2x0} \tag{55}$$

whence

$$A_m = A_a - A_b = \Delta_{340} - \Delta_{120} = \Delta_{34y} - \Delta_{xy0} \tag{56}$$

or,

$$A_m = \frac{r^2}{2} (\sin \alpha - \sin \beta) \tag{57}$$

here reading the absolute values of the angles α and β.

It is readily seen that

$$\phi = \frac{\beta - \alpha}{2} \tag{58}$$

and

$$\phi = \frac{180°}{u/t} . \tag{59}$$

We are concerned here with the particular case where $A_m = 0°$, or, from Eq. (57), $\sin \alpha = \sin \beta$. This can occur only when:

$$\alpha = 180° - \beta \tag{60}$$

whence:

$$\phi = \beta - 90° = 90° - \alpha \tag{61}$$

and, from Eqs. (36b) and (36c):

$$4 r^2 = d_a^{*2} + d_b^{*2}. \tag{62}$$

From (36a)

$$\sin^2 \frac{\alpha}{2} = \frac{1 - \cos \alpha}{2} = \frac{d_a^{*2}}{4r^2} \tag{63a}$$

$$\cos \alpha = \frac{2 r^2 - d_a^{*2}}{2 r^2} \tag{63b}$$

whence:

$$\sin \phi = \frac{d_b^{*2} - d_a^{*2}}{d_a^{*2} + d_b^{*2}} \tag{63c}$$

and

$$\cos \phi = \frac{2 \cos (\gamma_a/2) \cos (\gamma_b/2)}{\cos^2(\gamma_a/2) + \cos^2(\gamma_b/a)} \tag{63d}$$

whence:

$$\cos (\gamma_a/2) = \cos (\gamma_b/2) \frac{(1 + \sin \phi)}{\cos \phi}. \tag{63e}$$

This, with the helix equation (Eq. (29)) allows a determination of values for the two conformation angles for the case where $A_m = 0$ (using successive approximations). These values are shown in Table 8.

Table 8. Conformational angles for amphiverse helices where $A_m = 0$

γ_a	γ_b
0°	180°
20	171.54
40	163.75
60	157.25
80	152.60
100	150.40
120	151.20
140	155.60
160	164.55
180	180

11. Radial Helices

In those helices (11, 17) where each bond a passes through the helix axis, the area subtended by the bond a:

$$A_a = \frac{d_a^* [d_b^* - d_a^* \cos \phi]}{4 \sin \phi} = 0 \tag{64a}$$

whence, since $d_a^* \neq 0$

$$\cos \phi = \frac{d_b^*}{d_a^*} = \frac{\cos (\gamma_a/2)}{\cos (\gamma_b/2)}. \tag{64b}$$

It is readily shown that:

$$\tan (\gamma_a/2) = \frac{7 + \cos \gamma_b}{3 \sin \gamma_b}. \tag{64c}$$

Points defining this important boundary line between *amphiverse* and *star* helices are shown in Table 9. The values shown in 17 were obtained from these two equations, letting $\phi = 45°$.

Table 9. Radial helices

γ_a	γ_b
0°	180°
15	168.87
30	158.41
45	149.22
60	141.79
75	136.48
90	133.60
105	133.48
120	136.43
135	142.74
150	152.52
165	165.34
180	180

12. Step *(trans)* Helices

When the s-conformation of each bond a is *trans* ($\gamma_a = 180°$) then each bond b will be parallel to the helix axis. The resulting helix has a step-like structure as seen from the side (see Fig. 28). Its cross section is a regular polygon when u/t is an integer. Each *motif* includes one corner and one edge of that polygon, whence:
the number of sides:

$$n = u/t \tag{65}$$

the length of each side:

$$d^* = d \sin \theta = \frac{2\sqrt{2}}{3} d . \tag{66}$$

The external angle, *which does not correspond to* ϕ as that angle is denoted in other helices, is the conformation angle γ_b, whence:

$$\gamma_b = \frac{n-2}{n} 180° = \frac{(u/t - 2)}{u/t} 180° \tag{67a}$$

or

$$u/t = \frac{360°}{180° - \gamma_b} . \tag{67b}$$

It can be shown that the area subtended by each *motif*:

$$A_m = \frac{d^{*2}}{4} \tan (\gamma_b/2) = \frac{2\,d^2}{9} \tan (\gamma_b/2) \tag{68a}$$

and the cross section area:

$$A_t = \frac{80° \; d^2}{180° - \gamma_b} \; \tan \left(\gamma_b / 2 \right).$$ (68 b)

Table 10. Conformation angles for step helices of particular symmetry

u/t	γ	H
2	0°	0°
3	60	0.23271
4	90	0.30230
5	108	0.33286
6	120	0.34907
7	128.57	0.35869
8	135	0.36491
9	140	0.36914
10	144	0.37215

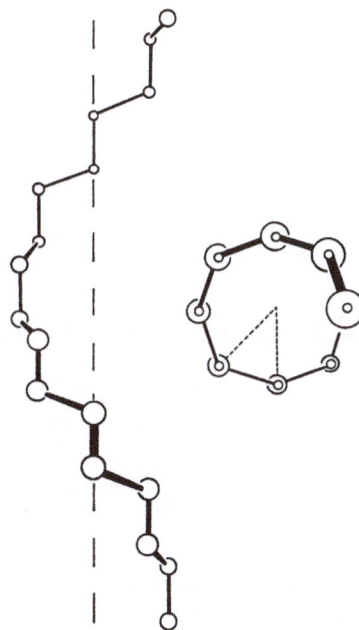

Fig. 28. A step helix with $u/t = 8/1$ ($\gamma_a = 180°$, $\gamma_b = 135°$), as seen from the side and in cross section

The rise for each *motif*:

$$P_m = d(1 - \cos \theta) = \tfrac{4}{3} d \qquad (69\,\text{a})$$

whence

$$L_t = \frac{480° \, d}{180° - \gamma} \qquad (69\,\text{b})$$

while

$$D_t = \frac{720° \, d}{180° - \gamma}. \qquad (70)$$

Hence:

$$H = \frac{2\pi}{3\sqrt{3}} \frac{\tan (\gamma_b/2)}{u/t}. \qquad (71)$$

The 3/1 step helices are important as the least strained of the all-staggered conformation available to isotactic poly-α-olefins.

Table 11. Helicity contour
points for step helices
($\gamma_a = 180°$)

H	γ_b
0.05	10.27°
0.10	21.35
0.15	34.26
0.20	49.06
0.25	66.84
0.30	88.62
0.35	120.88
0.375	148.58

V. On the Optical Activity of Polymethylene Chains

In a polymer sample in which all molecules have the same molecular weight and each has the same number, t, of identical helical turns:

$$[M]_D = t[M_t]_D = [\alpha]_D \frac{MW}{100} = [\alpha]_D \cdot \frac{t \cdot RW \cdot u/t}{100} \qquad (72)$$

where:

$[M_t]_D$ is the molecular rotatory contribution of each *turn unit*
$[\alpha]_D$ is the specific rotation
MW is the molecular weight
RW is the residue molecular weight, the contribution of each unit, u.

whence, the *residue rotation*:

$$[M_R]_D = \frac{[M_t]_D}{u/t} = [\alpha]_D \frac{RW}{100}. \tag{73}$$

This expression is independent of molecular chain length and so is suitable for use with polymers of mixed molecular weight. The *turn* molecular rotation contribution can be obtained from either of the models for optical rotation we have presented [12,14], either as a sum of contributions from four-atom units or by use of helical conductor equation (Eq. 1):

$$[M_t]_D = 652 \frac{L_t \cdot A_t}{D_t^2} \cdot R_t \, f(n). \tag{74}$$

In the case of a [2]-helical polyethylene chain:

$$R_t = u/t \cdot Ru \tag{75}$$

where Ru is the unit refraction, the contribution of two saturated C—C bounds (2.592). Since, in such helices:

$$D_t = 2 \times 1.54 \, u/t \tag{76}$$

and since the refractive index of a solution can be approximated as $\sqrt{2}$, $(f(n)) = \frac{[n^2 + 2]^2}{9\,n} = 1.2571$):

$$[M_R]_D = 200 \cdot H \cdot u/t. \tag{77a}$$

This equation leads to a *reductio ad absurdum* that may provide a significant refinement of the helical conductor model. It will be noted that as the step helices ($\gamma_a = 180°$) approach the absolutely planar zigzag chain structure ($\gamma_a = \gamma_b = 180°$) they acquire very large cross sections. The index of helicity approaches a constant value of about 0.385 but the expected residue rotations approach infinity. (See Table 12). This result is, at least intuitively, absurd.

It is our present view that this absurdity arises from any or all of several assumptions that entered into the derivation of the helical conductor equation (Eq. 1): [12]

a) It was assumed that the helix was small relative to the wavelength of light and, thus, at any one instant completely bathed in an electric field that would be uniform in space (though not in time). A very large helix, then, would not experience such a uniform field and so not obey the helical conductor equation except for very long wavelength radiation.

Table 12. Calculated residue rotation for step helices ($\gamma_a = 180°$) as γ_b approaches 180°

γ_b	ϕ	u/t	H	$[M_R]$
178°	1	180	0.38486	13,860
179	0.5	360	0.38489	27,712
179.9	0.05	3,600	0.38490	277,128
179.99	0.005	36,000	0.38490	2,771,280

We note that macroscopic helices do, indeed, give rotation of radio frequency radiation. [10,11] The constant for the helical conductor equation contains the term: $\lambda_0^2/(\lambda^2 - \lambda_0^2)$ and would decrease with increasing wavelength. If, then, the light had a long enough wavelength for the equation to apply, the numerical constant would be much smaller and, so, the calculated rotation. If this is the sole factor involved in the absurdity no change in Eq. (1) is required for use with small molecules.

b) It was assumed that the chain of bonds is a perfect conductor. A damping factor may be required if this is not exactly correct. Indeed, the numerical values calculated for relatively short chains (see the treatment of twistane [16]) are disturbingly high and suggest the possibility that such a factor might be required in *all* applications of the equation. It would not, in all probability, have a *major* effect on the utility of the helical conductor model, as used in small molecules.

c) It was assumed in the derivation of Eq. (1) that *all* of the electrons in a chain are perturbed by light and, thus, that the electric moment (m) induced by the electric field of the light (E) is related to the *total* polarizability (α) of the system:

$$m = \alpha E. \qquad \text{(Eq. (26) of [12])}$$

If, instead, we assume that only one bond is perturbed at any one time then it would be the *average* polarizability that is important. This would lead to the use of the *average* bond refraction ($\overline{R}_\lambda$) rather than the *sum* of bond refractions in Eq. (1).

$$[M]_D = 652 \frac{LA}{D^2} \overline{R}_\lambda f(n). \qquad (1\,\text{b})$$

If so, the right hand terms of Eqs. (74) and (77) should be divided by $2 \times u/t$, whence, for [2]-helices:

$$[M_R]_D = 100 \cdot H. \qquad (77\,\text{b})$$

The conformation map shown in Fig. 26 then becomes (multiplying contour values by 100) a map of the predicted rotatory contributions of the *backbones* of [2]-helices of vinyl polymers as a function of conformation angle. The step helices now approach a maximum residue rotation of 38.5 — the factors discussed above would reduce this to zero for very large helices.

These considerations, thus, lay the groundwork for tests among several semi-empirical approaches to the estimation of optical rotation of bond systems regarded as helices. Should it be necessary to use Eq. (1b) rather than (1a), then a sweeping reassessment of the use of the helical conductor model will be required. However that test turns out, a test between that model and the simple conformational dissymmetry model becomes possible on the basis of the material shown in Table 1. At this point it should be said that our calculations on twistane [16] support the helical conductor model but that the results obtained by Pino and his co-workers [17,18] on the chiroptical properties of isotactic polymers prepared from chiral α-olefins support the conformational dissymmetry model. [We are not able, at present anyhow, to account for their results with the helical conductor model].

Thus the story is not done.

VI. References

[1] Fresnel, A.: Mem. Acad. Sci. *7*, 45 (1827).
[2] van't Hoff, J.: Bull. Soc. Chim. France, *23*, 295 (1875).
[3] le Bel, A.: Bull. Soc. Chim. France *22*, 337 (1874).
[4] Gibbs, J. W.: Am. J. Sci. (3), *23*, 460 (1882).
[5] Drude, P.: Göttinger Nachrichten *1892*, 366; Theory of Optics (trans. by C. R. Mann and R. A. Millikan) (publ. 1900), Dover reprint 1959, pp. 400—407.
[6] Rosenfeld, L.: Z. Physik *52*, 161 (1928).
[7] Kuhn, W.: Z. Physik. Chem. *4B*, 14 (1929); — Kuhn, W.: Stereochemie (ed. K. Freudenberg, pp. 317—434. Leipzig: Deuticke 1933.
[8] Kirkwood, J. G.: J. Chem. Phys. *5*, 479 (1937). — Wood, W. W., Fickett, W., Kirkwood, J. G.: J. Chem. Phys. *20*, 561 (1952).
[9] Kauzmann, W.: Quantum chemistry, pp. 616—636, 703—725. New York: Academic Press 1957
[10] Winkler, M. H.: J. Phys. Chem. *60*, 1656 (1956); *62*, 1342 (1958).
[11] Tinoco, Jr., I., Freeman, M. P.: J. Phys. Chem. *61*, 1196 (1957).
[12] Brewster, J. H.: Topics stereochem. *2*, 1 (1967).
[13] Lardicci, L., Menicagli, R., Caporusso, A. M., Giacomelli, G.: Chem. Ind. (London) *1973*, 184.
[14] Brewster, J. H.: J. Am. Chem. Soc. *81*, 5475 (1959).
[15] Cahn, R. S., Ingold, C. K., Prelog, V.: Angew. Chem., Intern. Ed. Engl. *5*, 385 (1966).
[16] Brewster, J. H.: Tetrahedron Letters *1972*, 4355.
[17] Pino, P.: Advan. Polymer Sci. *4*, 393 (1965).
[18] Pino, P., Ciardelli, F., Zandomeneghi, M.: Ann. Rev. Phys. Chem. *21*, 561 (1970).

Received August 24, 1973

Recent Aspects of Cyclopropanone Chemistry

Prof. Harry H. Wasserman, Dr. George M. Clark
and Dr. Patricia C. Turley

Department of Chemistry, Yale University, New Haven, Connecticut, USA

Contents

H. H. Wasserman, G. M. Clark, and P. C. Turley

1. Introduction

Cyclopropanones are highly reactive organic systems containing a number of labile sites on a small carbon skeleton. They represent valuable substrates for studying theoretical aspects of the chemistry of small strained ring systems and are of special interest in synthesis because of the variety of transformations in which they take part.

Until quite recently, cyclopropanones were known only as transient intermediates, or in the form of derivatives such as the hydrate or hemiacetal. [1] During the past decade, however, through the work of the groups at Columbia [2] and Amsterdam [3] among others, methods have been developed for preparing a number of representatives of this class. Table 1 lists various cyclopropanones which have been isolated as well as several, more elusive examples which have been characterized in solution.

A significant step in studying the chemistry of cyclopropanones has resulted from the discovery that many labile carbonyl derivatives such as hemiacetals and carbinol amines are useful precursors of the parent ketone. [4-6] Such derivatives may be isolated, purified and used as cyclopropanone substitutes or, alternatively, may be generated in solution and used as *in situ* precursors. As a result of these advances, exploration of cyclopropanone chemistry has recently been accelerated. The aim of this article is to review some of this chemistry, noting areas where there may be potential applications in synthesis.

In reviewing the methods available for preparing cyclopropanones (Section 2), it becomes clear that a reliable, general method for the preparation of three-membered ketones has still to be devised. For example, the key combination of X and Y substituents and conditions needed to effect a conversion such as shown in Scheme 1 has not yet been discovered.

In our discussion of cyclopropanone chemistry, much of the attention is devoted to derivatives of type *1* where X and Y are labile groups

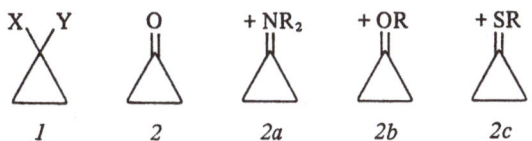

H. H. Wasserman, G. M. Clark, and P. C. Turley

Table 1

Cyclopropanone	Method of preparation	Solvent	Temp., °C	Yield, %	State in which studied	Method of characterization	Ref.
	$CH_2N_2 + CH_2C=O$	CH_2Cl_2	−78	50—60	Solution	IR, NMR, UV, reactions	2)
	$CH_2N_2 + CH_2C=O$	$CHCl_3$; propane; $CFCl_3$	−78	50	Impure liquid, stable few days at liq N_2 temp.	IR, NMR, reactions	3)
	$CH_2N_2 + CH_2C=O$	$CFCl_3$	−78	—	Gas phase mixture	microwave	63)
	$CH_2N_2 + CH_2C=O$	N_2 (solid)	−253	—	N_2 matrix; gas phase mixture	IR	127)
	$CH_3CHN_2 + CH_2C=O$	CH_2Cl_2	−78	—	0.4 M solution	IR, NMR	10)
	$CH_2N_2 + {>}{=}C=O$	CH_2Cl_2	−78	>93	Solution	IR, NMR, UV, reactions	103)
, hv		—	—	"Low"	—	IR	10)
	$CH_3CHN_2 + {>}{=}C=O$	CH_2Cl_2	−78	>60	Solution	IR, NMR, reactions	96)
		Pentane	36	—	Solution	IR, ractions	8)
		CH_2Cl_2	0	—	Solution	IR, NMR, UV, reactions	10)
, hv		N_2 (solid)	−269	—	Matrix	IR	128)
		Cyclohexane	24	—	Solution	IR	128)

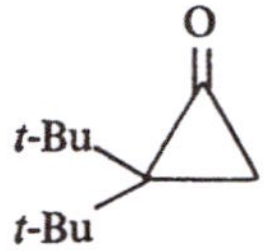

t-Bu, t-Bu cyclopropanone	$\begin{array}{c}R\\R\end{array}$C=C=CH$_2$, CH$_3$COOH	CH$_2$Cl$_2$	—	—	Solid, mp 41–43 °C	Analysis, IR, NMR, UV mass spec., reactions 55a)
H, t-Bu cyclopropanone	RCHBrCOCH$_2$R, base	—	—	20–40	Solid, mp 24–26 °C	Analysis, IR, NMR, UV, mass spec., reactions 13)
epoxide (R, H)	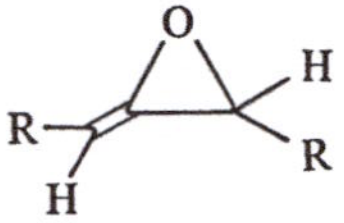	—	100	—	—	IR, UV 51)
X-shaped diketone	Oxidation of corresp. alcohol	DMSO	—	—	Solid, mp 240–260 °C, d	IR, NMR, mass spec., reactions 33)
	>N=C=N<, DMSO	—	—	—	—	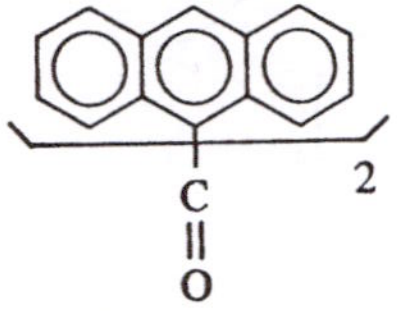 — 32)
anthracene C=O, hv (2)		—	—	100%	—	—
Ph—O—Ph furan + cyclopropanone		—	—	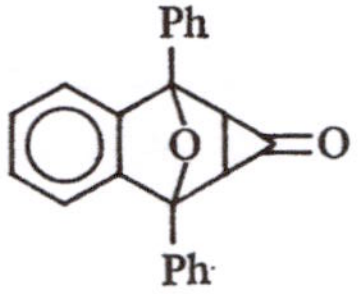 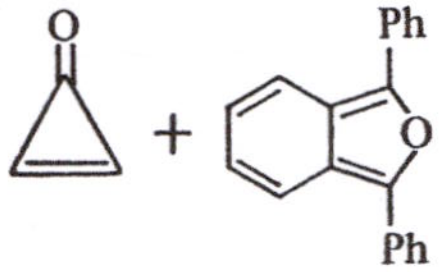 Solid, mp 151–153 °C	R, NMR, mass spec., methanol addition 62)

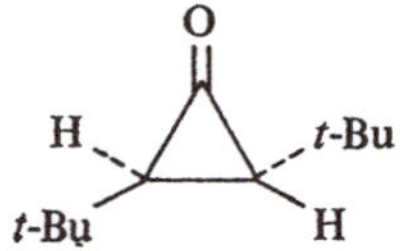

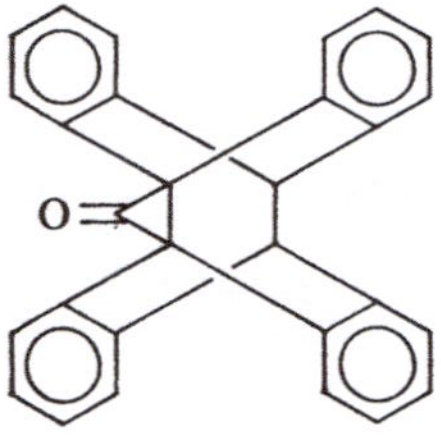

77

H. H. Wasserman, G. M. Clark, and P. C. Turley

Table 1 (continued)

Cyclopropanone	Method of preparation	Solvent	Temp., °C	Yield, %	State in which studied	Method of characterization	Ref.
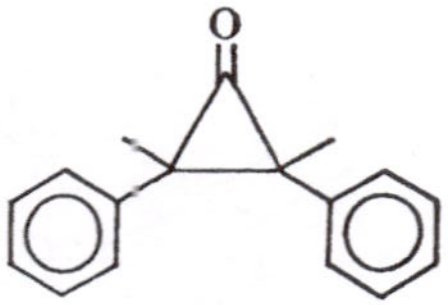 [diphenyl cyclopropanone structure]	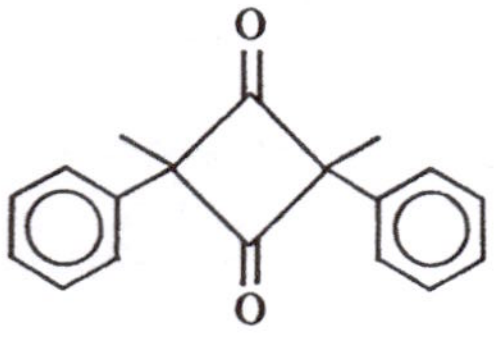 [diphenyl cyclobutanedione structure], hν	—	—	"Low"	—	IR	10)
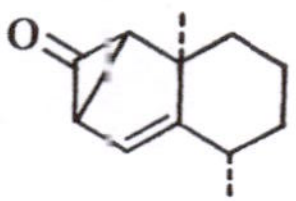 [bicyclic structure]	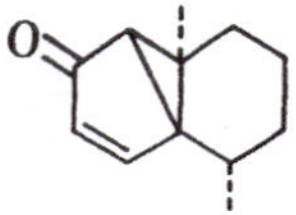 [bicyclic structure], hν	Film	−190	—	—	Decomposes −115 °C IR, reactions	34)
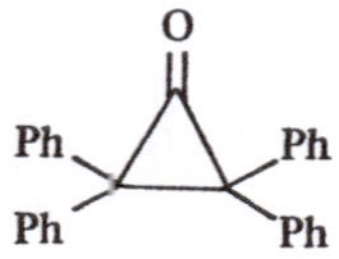 [tetraphenyl cyclopropanone structure]	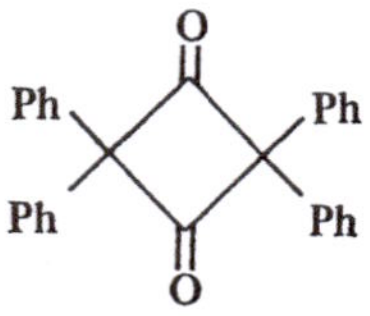 [tetraphenyl cyclobutanedione structure], hν	—	—	"Low"	—	IR	10)

(usually other than halogen) which may be converted to *2* or derivatives of *2a-c* under suitable reaction conditions. 1,1-Dihalocyclopropanes are mentioned only as intermediates in the formation of more labile systems corresponding to *1*. We have also omitted discussions of the preparation and chemistry of cyclopropenones.

2. Preparative Methods

Cyclopropanones have been synthesized by a variety of methods, the most straightforward of which involves the addition of carbenes or carbenoids[a], *e.g.* CH_2N_2 [1-3] or CH_2I_2 [7], to a double bond. Other, less general routes include photochemical decarbonylation of cyclo-butane-1,3-diones [8-10], ring closure of 1,3-disubstituted ketones or ketals [11,12] and the Favorskii reaction.[13] While many of these cyclopropanones are too unstable to be isolated, they may be trapped in the form of hemiacetals, carbinol amines or other carbonyl addition products, and regenerated under suitable reaction conditions.[4-6] A particularly useful cyclopropanone source is the acetic acid adduct, 1-acetoxycyclo-propanol.[5] This derivative is easily purified by distillation and readily undergoes substitution at the 1-position by nucleophilic reagents.[14]

2.1. Diazoalkane-Ketene Reaction

A versatile synthesis of cyclopropanones and closely related derivatives is provided by the diazoalkane-ketene reaction as shown in Scheme 2. Using this method, the parent ketone [2,3] and alkyl-substituted cyclopropanones [10] have been prepared in yields of 60—90% based upon the concentration of diazoalkane[b] (Table 2). The reaction is rapid at Dry Ice-acetone temperatures and is accompanied by evolution of nitrogen. Although most cyclopropanones are not isolable, dilute solutions of *3* (0.5—0.8 M) may be stored at -78 °C for several days or at room temperature in the presence of suitable stabilizing agents.[15] The hydrate and hemiketal derivatives are readily prepared by the addition of water or alcohols to the solutions of *3*.[2,3,5]

---

[a] The term "carbenoid" refers to species which may *not* be free divalent carbon intermediates but exhibit carbene-like properties (Closs, G. L., Moss, R. A.: J. Am. Chem. Soc. *86*, 4042 (1964)).

[b] To minimize formation of cyclobutanone, the concentration of the diazoalkane is kept at one-half that of the ketene.

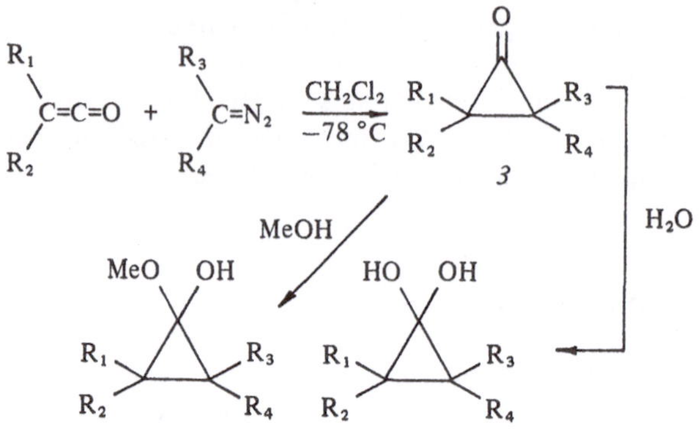

Scheme 2

As shown in Table 2, the application of this method to the synthesis of aryl-substituted cyclopropanones [16] and cyclopropanone acetals [17,18] has been moderately successful, although products other than the expected ketones may be obtained. For example, the oxadiazoline 4 and not tetraphenylcyclopropanone is formed when diphenylketene is allowed to react with diphenyldiazomethane.[19]

4

Similarly, cyclopropanones have not been found among the products from the reaction of ethyl diazoacetate with dimethylketene or diphenylketene although cyclopropanone intermediates may be involved (see Section 4.1.5).[20] Attempts to prepare the cyclic thioketal 5[c] by the addition of diazomethane to the trithiane 6 were also unsuccessful.[21]

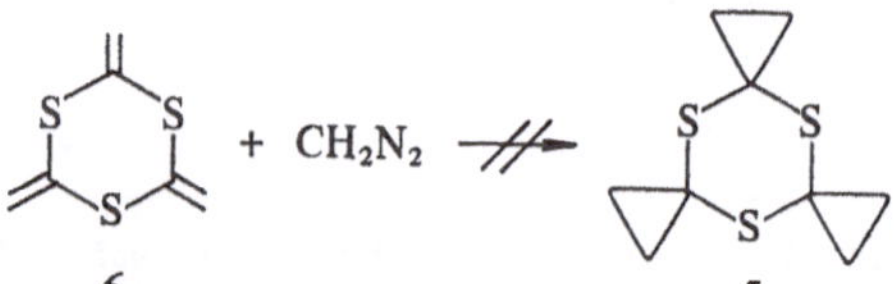

6   5

---

c) See Section 2.4 for a preparation of 5.

Table 2. Cyclopropanones by the diazoalkane-ketene reaction

Ketene	Diazoalkane	Products	Yield, (%)	Conditions	Ref.
$H_2C=C=O$	CH_2N_2		50–75 / 50	CH_2Cl_2 $-78\,°C$ / $CHCl_3$ $-78\,°C$	2,133) / 3)
$H_2C=C=O$	CH_3CHN_2			CH_2Cl_2 $-78\,°C$	10)
$(CH_3)_2C=C=O$	CH_2N_2	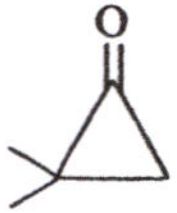	>93	CH_2Cl_2 $-78\,°C$	10)
$(CH_3)_2C=C=O$	CH_3CHN_2	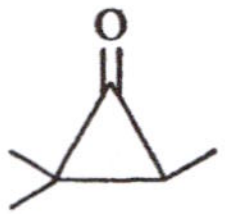	>60	CH_2Cl_2 $-78\,°C$	96)
$H_2C=C=O$	$ArCHN_2$	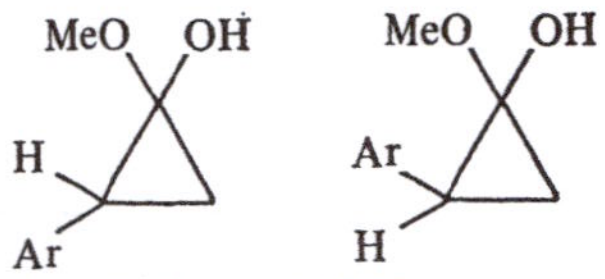 $Ar = C_6H_5, p-ClC_6H_4$		CH_3OH	16)

81

Table 2. (continued)

Ketene	Diazoalkane	Products	Yield(%)	Conditions	Ref.
$H_2C=C=O$	$Ar-C(N_2)-CH_3$	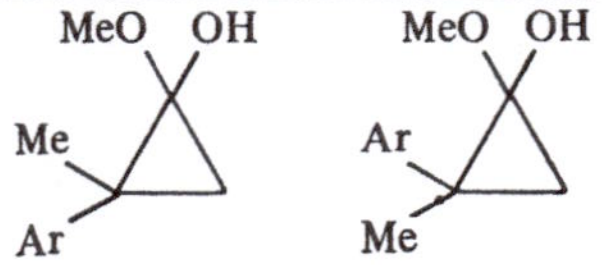 (cyclopropane structures) MeO OH / Me Ar and MeO OH / Ar Me; Ar $= C_6H_5$, p-ClC_6H_4		CH_3OH	16)
$H_2C=C(OEt)_2$	CH_2N_2	 (cyclopropane) EtO OEt	40	CuBr catalyst	17)
$PhCH=C(OEt)_2$	CH_2N_2	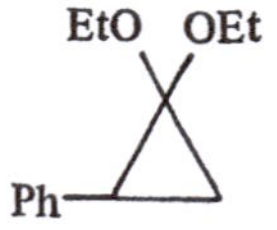 (cyclopropane) EtO OEt / Ph	12	CuBr catalyst	17)
$H_2C=C(OEt)_2$	$PhCHN_2$	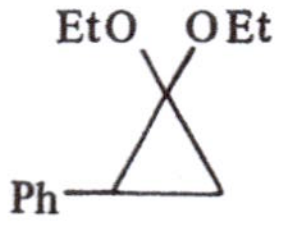 (cyclopropane) EtO OEt / Ph	82	CuBr catalyst	17)
$(CH_3)_2C=C(OMe)_2$	$N_2CHCOOEt$	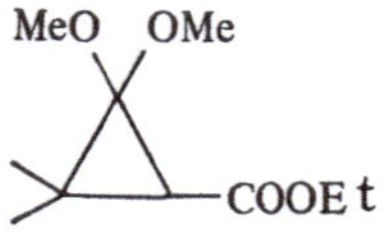 (cyclopropane) MeO OMe / COOEt		Cu-Bronze	18)

2.2. Other Carbene or Carbenoid Reactions with Olefins

The reaction of carbenes with appropriately substituted olefins provides a useful method for the preparation of many cyclopropanone derivatives. The Simmons-Smith procedure [22] and reactions involving base-generated carbenes, e.g. $CHCl_3/KO$-t-Bu, are particularly useful.

The esters, 1-ethoxycyclopropyl acetate (7a) and benzoate (7b) have been synthesized by the addition of the Simmons-Smith reagent [22] to 1-ethoxyvinyl acetate and benzoate, respectively. [4] A potential difficulty in this reaction lies in the fact that zinc iodide, a Lewis acid, is generated in the process and may induce cyclopropane ring opening (Section 4.3.3). However, when glyme is used as a solvent, the acid-labile bonds remain intact since the zinc salt is insoluble in this medium. [23]

$$H_2C=C \begin{smallmatrix} OEt \\ OCOR \end{smallmatrix} + CH_2I_2, \ Zn/Cu \ \xrightarrow{\text{glyme}} $$

EtO OCOR

7   a, R = Me
     b, R = Ph

The Simmons-Smith reagent also adds to ketene acetals forming cyclopropanone acetals, as shown in the formation of 1,1-dimethoxy-2,2-dimethylcyclopropane (8) from 1,1-dimethoxy-2,2-dimethylethylene. [17] Similarly, with methylene iodide, ketene o-xylylene acetal (9) affords the corresponding cyclopropanone derivative (10) (70%). [24]

$$\begin{smallmatrix} Me \\ \\ Me \end{smallmatrix} C=C \begin{smallmatrix} OMe \\ \\ OMe \end{smallmatrix} + CH_2I_2, \ Zn/Cu \ \longrightarrow $$

MeO OMe

8

$$H_2C=C \begin{smallmatrix} O \\ O \end{smallmatrix} + CH_2I_2, \ Zn/Cu \ \longrightarrow$$

9   10

Cyclopropanone acetals may also be prepared by the addition of other one-carbon species to ketene acetals as shown in Table 3. Thus, McElvain and Weyna have synthesized several cyclopropanone derivatives in good yield by employing dichlorocarbene and chlorophenyl-carbene.[25] However, under the strongly alkaline conditions employed in generating these carbenes, substituted cyclopropanone acetals may react further. For example, the reaction of dichlorocarbene and phenyl-ketene dimethylacetal affords *11*, the *ortho* ester of 3-phenylpropynoic acid, rather than the cyclic acetal, presumably by the route shown below (Scheme 3).[25]

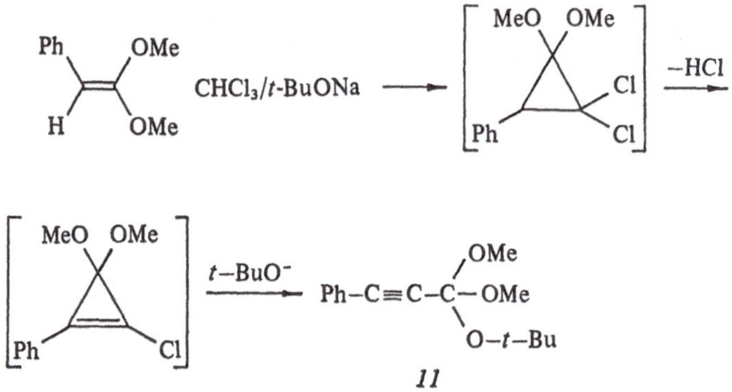

Scheme 3

Schöllkopf and co-workers have synthesized a number of cyclo-propanone acetals by the addition of various sulfur- and oxygen-con-taining carbenes to ketene diethylacetals (Table 3).[26,27] Similarly, cyclopropanone dithioacetals may be prepared by the addition of the *bis*-thiomethyl and *bis*-thiobenzylcarbenes *12a, b* to olefins.[29] However, cyclopropanone acetal formation by this method requires double bonds with considerable electron enrichment and the yields are generally low. With unsubstituted olefins such as cyclohexene, the carbenes *12a, b* tend to form dimeric and trimeric products such as *13* and *14*, instead of the double bond addition products.

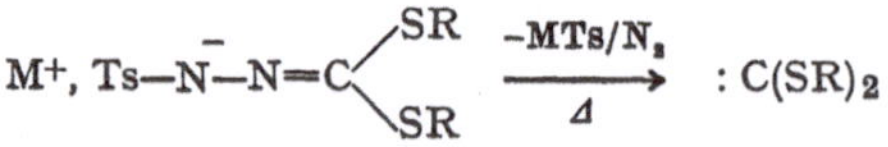

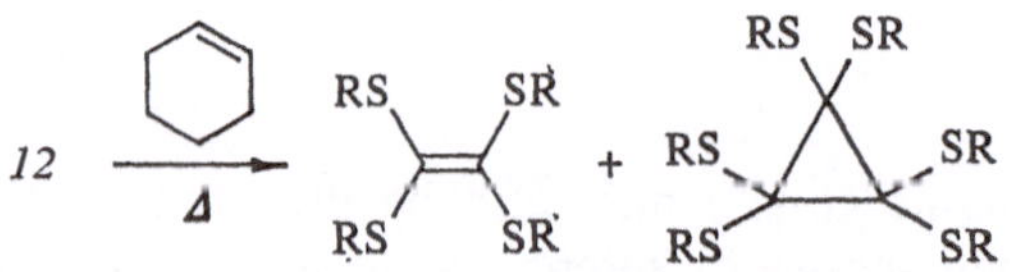

Table 3. Cyclopropanone acetals and thioacetals from carbene additions

Ketene	Carbene source	Cyclopropanone acetal	% Yield	Ref.
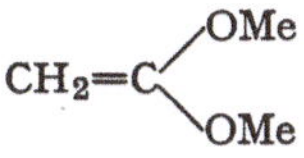 $CH_2=C$ (OMe)(OMe)	$CHCl_3$, NaO-t-Bu	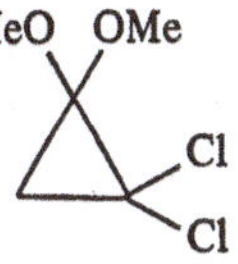 MeO OMe / Cl / Cl	40	25)
	$CHCl_3$, NaOH		62	129)
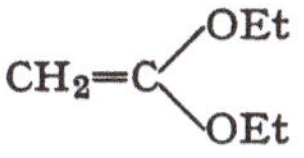 $CH_2=C$ (OEt)(OEt)	$CHCl_3$, NaO-t-Bu	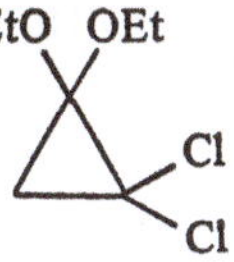 EtO OEt / Cl / Cl	68	25)
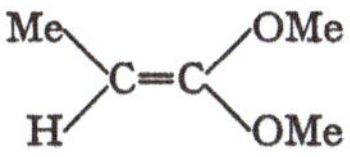 Me, H, $C=C$ (OMe)(OMe)	$CHCl_3$, NaO-t-Bu	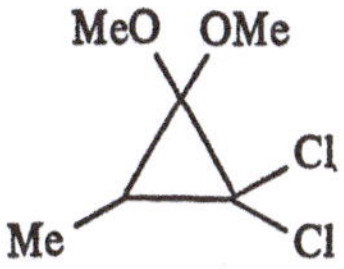 MeO OMe / Cl / Me Cl	52	25)
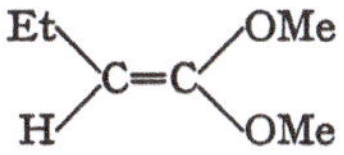 Et, H, $C=C$ (OMe)(OMe)	$CHCl_3$, NaO-t-Bu	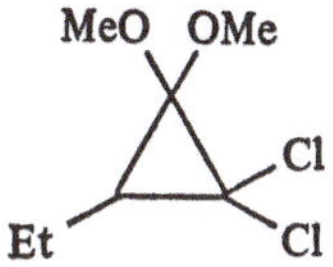 MeO OMe / Cl / Et Cl	56	25)
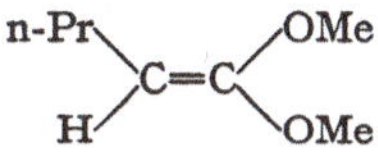 n-Pr, H, $C=C$ (OMe)(OMe)	$CHCl_3$, NaO-t-Bu	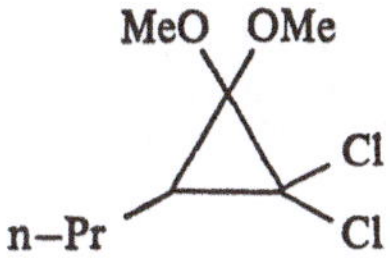 MeO OMe / Cl / n–Pr Cl	55	25)
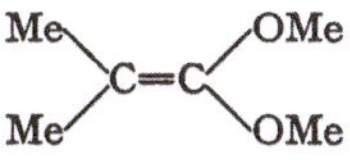 Me, Me, $C=C$ (OMe)(OMe)	$CHCl_3$, NaO-t-Bu	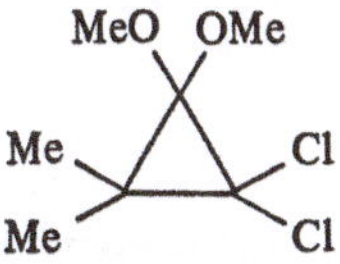 MeO OMe / Cl / Me Me Cl	61	25)
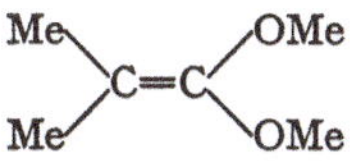 Me, Me, $C=C$ (OMe)(OMe)	$PhCHCl_2$, NaO-t-Bu	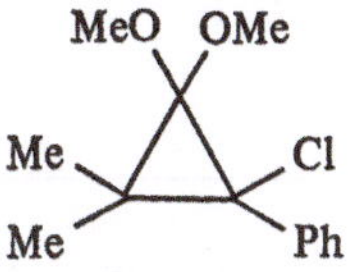 MeO OMe / Cl / Me Me Ph	77	25)
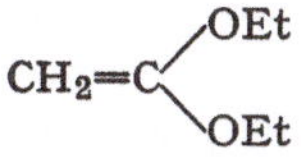 $CH_2=C$ (OEt)(OEt)	$ClCH_2SPh$, n-BuLi	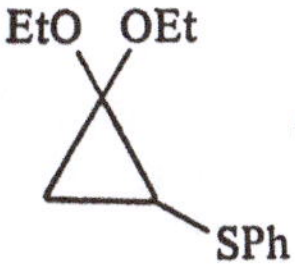 EtO OEt / SPh	80	26a,b)

Table 3 (continued)

Ketene	Carbene source	Cyclopropanone acetal	% Yield	Ref.
$\overset{H}{\underset{Ph}{}}C=C\overset{OEt}{\underset{OEt}{}}$	Cl_2CHSPh, KO-t-Bu	EtO OEt, triangle with SPh, Cl, Ph	—	26c)
$CH_2=C\overset{OEt}{\underset{OEt}{}}$	Cl_2CHSPh, KO-t-Bu	EtO OEt, triangle with SPh, Cl	40—60	26c)
$CH_2=C\overset{OEt}{\underset{OEt}{}}$	$ClCH_2OC_6H_5$, BuLi	EtO OEt, triangle with OPh	33	27)
$CH_2=C(OEt)_2$	$TsNH-N=C(SMe)_2$, NaH	EtO OEt, triangle with SMe, SMe	30	28)
morpholine-$N-\overset{}{\underset{Ph}{C}}=CH_2$	$TsNH-N=C(SMe)_2$, NaH $TsNH-N=C(SMe)_2$, t-BuOK	MeS SMe, triangle with N-morpholine, Ph	32 23	28) 28)
morpholine-$N-\overset{}{\underset{Ph}{C}}=CH_2$	$TsNH-N=C(SBz)_2$, NaH	BzS SBz, triangle with N-morpholine, Ph	8	28)
$CH_3CH=CH-OPr$ (cis)	$TsNH-N=C(SMe)_2$, NaH	Me, triangle with MeS, MeS, OPr	11	28)

In a related reaction, 1-chloro-1-thiomethyl [29] and 1-chloro-1-thiophenyl [26c] cyclopropanes (*15a, b*) have been synthesized by the addition of chlorothiomethyl or chlorothiophenylcarbene to olefins. With the exception of 1-chloro-1-phenylthio-2,2-dimethylcycloprop-anone (56—60%) [26c], the yields are generally low (8—28%) since carbene dimerization competes with cyclopropane formation.

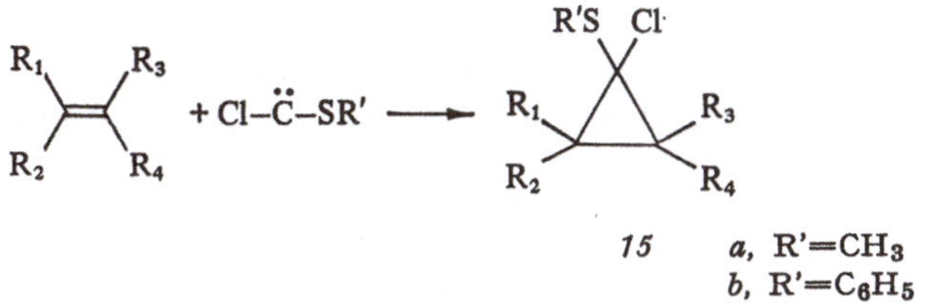

$$R_1R_2C=CR_3R_4 + Cl-\overset{..}{\underset{..}{C}}-SR' \longrightarrow \text{(cyclopropane with R'S, Cl, R}_1, R_2, R_3, R_4)$$

15   a, R'=CH₃
      b, R'=C₆H₅

Other carbene reactions include the preparation of 1-alkoxy-1-chlorocyclopropanes (16) by the treatment of α,α,α-trichloro ethers with methyllithium in the presence of lithium iodide.[30]

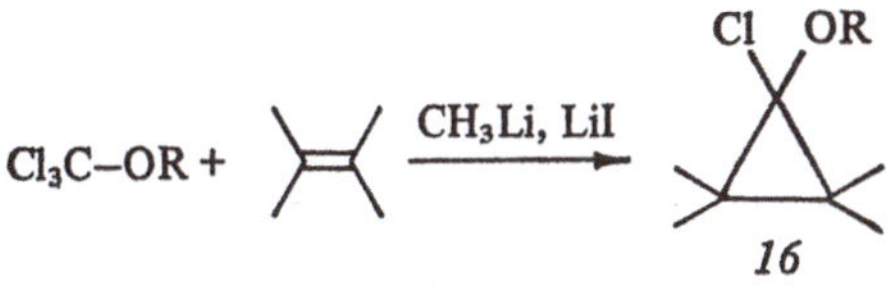

$$Cl_3C-OR + \text{(alkene)} \xrightarrow{CH_3Li, LiI} \text{(cyclopropane 16 with Cl, OR)}$$

16

2.3. Photochemical Methods

The preparation of cyclopropanones by photochemical decarbonylation [8-10] is illustrated by the formation of 1-ethoxycyclopropanol from the photolysis of tetramethylcyclobutane-1,3-dione (17) in ethanol.[9a] As shown in Scheme 4, the desired hemiacetal is accompanied by a number of side products including ethyl isobutyrate which results from the

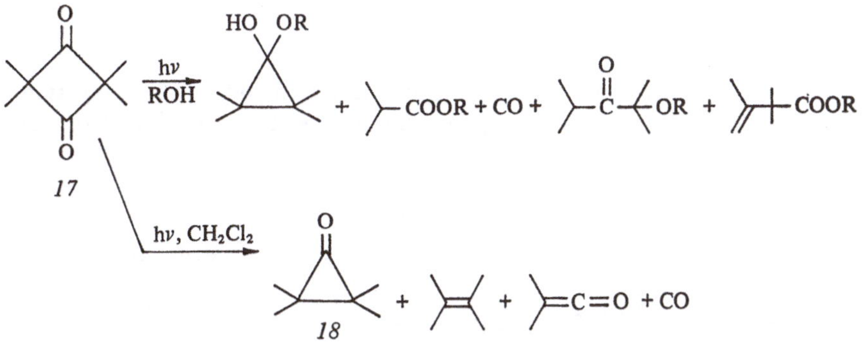

Scheme 4

addition of ethanol to dimethylketene. When photolysis of 17 is carried out in methylene chloride, tetramethylcyclopropanone (18) is obtained.[10] The low yield (10—20%) is due in part to a second stage decarbonylation to tetramethylethylene. When the cyclobutadiones *19—21* were irradiated [10], cyclopropanones were detected in the ir spectra of the reaction mixtures, but the yields were low and the photochemistry complex.

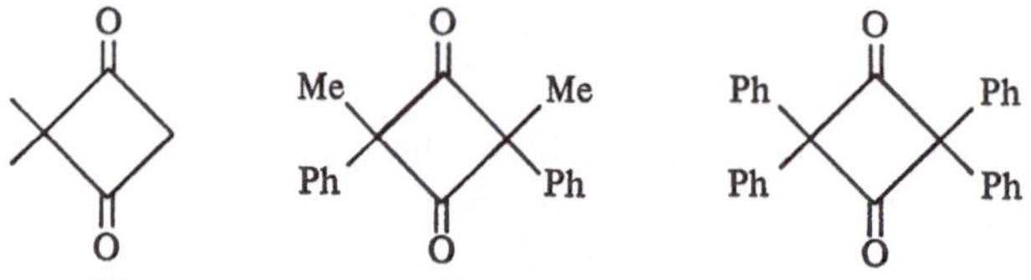

19   20   21

Hostettler has examined the photochemistry of several 3-substituted derivatives of 2,2,4,4-tetramethylcyclobutanones (Table 4).[31] In solvents containing proton donors, two major products were observed, the rearranged cyclic acetal *22* and the decarbonylation product *23*. In inert solvents such as ether, cyclohexane and benzene, formation of *22* did not occur.

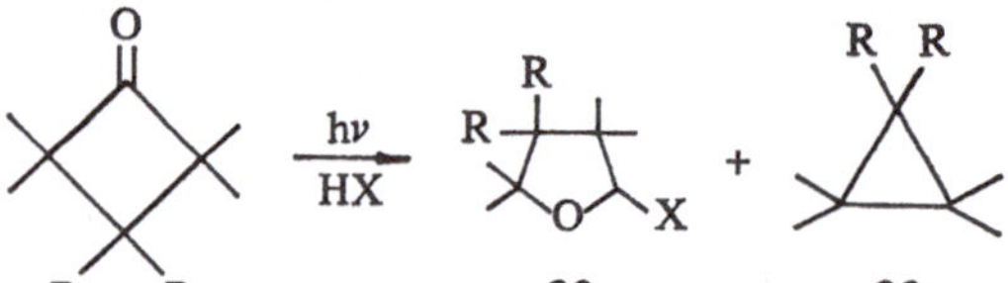

22   23

In cases where there is an exocyclic double bond at the 3-position, *e.g. 24*, reductive C_2—C_3 ring opening (*25 → 26*) under the work-up conditions (gas chromatography) has been observed.

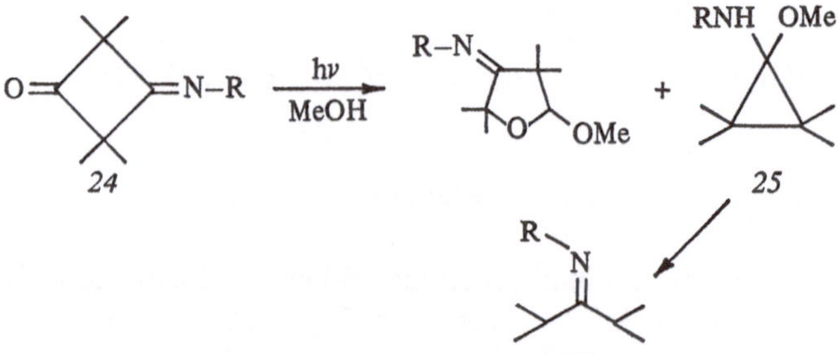

24   25

26

In other special cases, cyclopropanones have been prepared by photochemical rearrangements. For example, 9,9′-dianthracyl ketone (27) [32] and carbinol (28) [33] undergo ring closure to their respective

Table 4. Photolysis of 3-substituted cyclobutanones

Ketone	Solvent	Products (% yield)
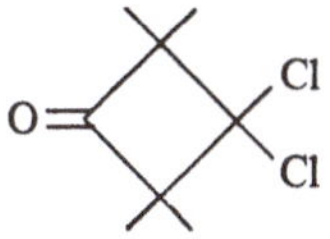 (3,3-dichloro tetramethylcyclobutanone)	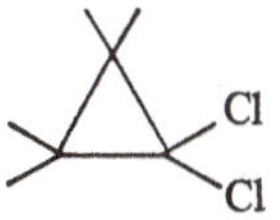 Cyclohexane	Cl, Cl cyclopropane (33)
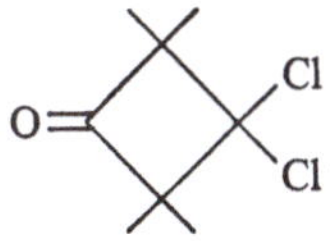 (3,3-dichloro tetramethylcyclobutanone)	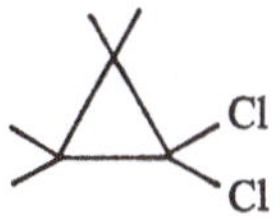 MeOH	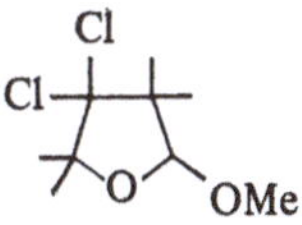 Cl, Cl cyclopropane (35) + Cl, Cl—OMe oxolane (30)
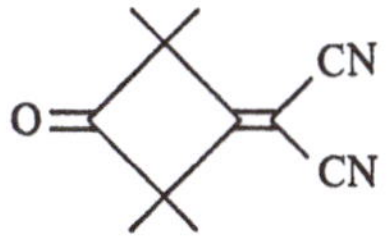 (dicyanomethylene tetramethylcyclobutanone)	Benzene	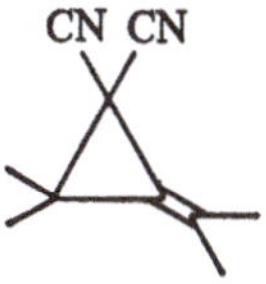 CN CN cyclopropane + NC CN alkene (35, total) 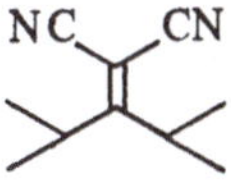
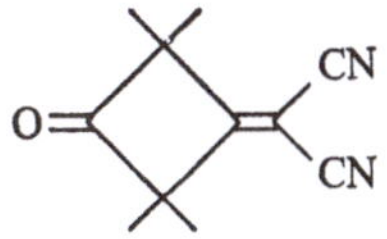 (dicyanomethylene tetramethylcyclobutanone)	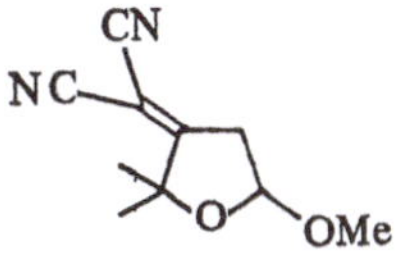 MeOH	NC, CN—OMe oxolane (57)
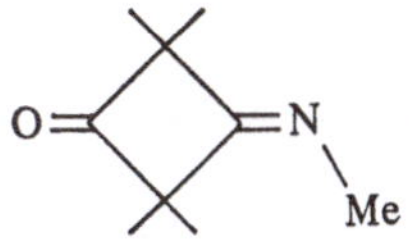 (N-Me imine tetramethylcyclobutanone)	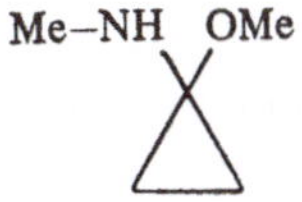 MeOH	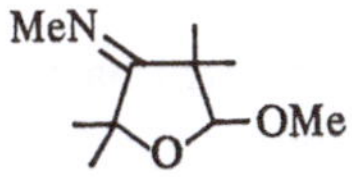 Me–NH OMe cyclopropane (23) + MeN OMe (73)
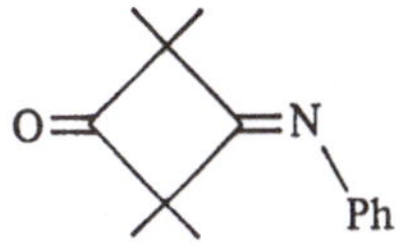 (N-Ph imine tetramethylcyclobutanone)	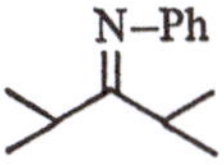 MeOH	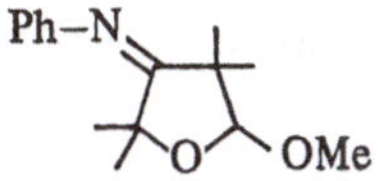 N–Ph (~ 20) + Ph–N OMe (~ 30)

cyclopropane derivatives 29 and 30. The cyclopropanol 30 may be converted to the ketone 29 by oxidation with manganese dioxide or with diisopropylcarbodiimide in dimethylsulfoxide.

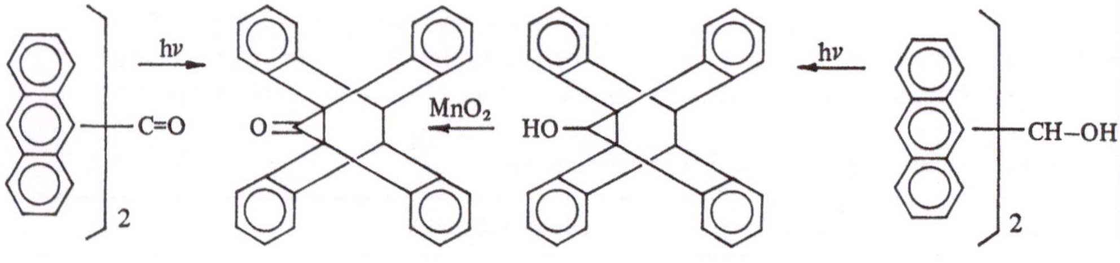

The cyclopropyl enone *31* was photoisomerized to the cyclopropanone *32* at liquid nitrogen temperatures.[34] The unstable *32* was characterized on the basis of ir spectra and thermal and photochemical reactions (see Section 4.1.5). When *31* was irradiated at room temperature in 45% acetic acid, the phenol *33* was produced, most probably through *32* and the spiro dienone intermediate *34* (Scheme 5).[35]

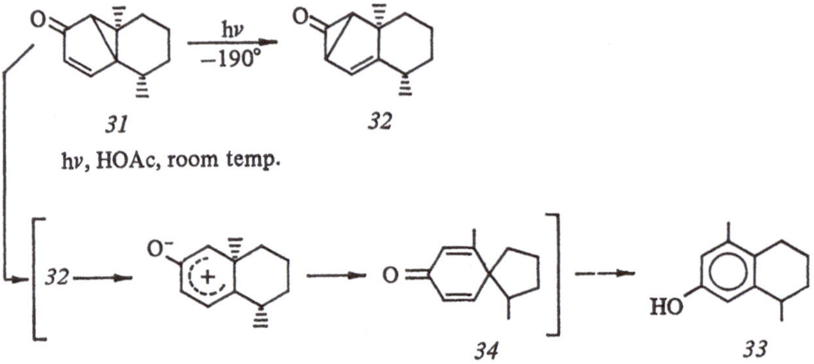

Scheme 5

Upon irradiation, the pyrrolinone *35* undergoes α-cleavage to give the ethoxycyclopropyl carbamate *36* (70%).[36] This reaction resembles that of 5,5-diphenyl and 5,5-dimethyl-2-cyclopentenone [37], which yields cyclopropane derivatives under conditions normally leading to 2-cyclopentenone dimers. [38].

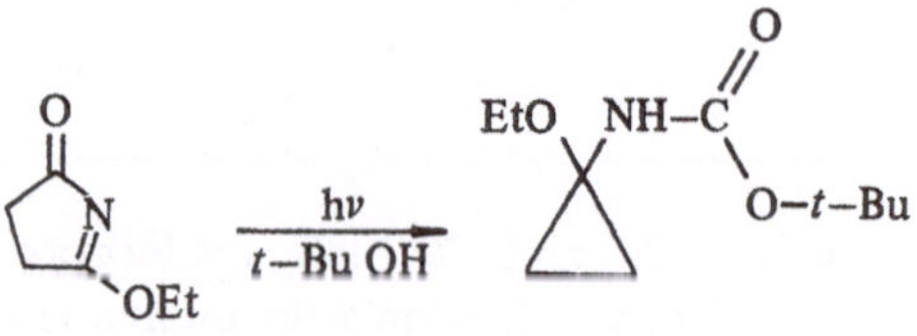

2.4. Dehalogenation and Dehydrohalogenation

Cyclization reactions by 1,3-dehalogenation or dehydrohalogenation[d] have provided additional routes to cyclopropanones. The application of the former reaction to the formation of 3-ring ketones was suggested by the isolation of the cycloadduct *37* from the reaction of α,α'-dibromo-benzyl ketone with sodium iodide in the presence of the trapping agent furan.[39] More recently, Giusti [11] has synthesized several cyclopropanone

ketals by treating the ketals of 1,3-dibromo-2-propanones with magnesium in tetrahydrofuran or with zinc in hexamethylphosphoric triamide (Table 5). Although allenes are formed as by-products, the yields of the cyclopropanone derivatives are generally satisfactory.

1,3-Dibromopropanones may also be reduced electrolytically to form cyclopropanone derivatives as illustrated by the formation [12] of 1-methoxytetramethylcyclopropanol, *38*, and 1-methoxy-2,2-dimethyl-cyclopropanol, *39*, in excellent yields.

38 $R_1 = R_2 = CH_3$
39 $R_1 = H; R_2 = CH_3$

A variety of cyclopropanone dithioketals have been prepared by the dehydrohalogenation of 3-halo dithioketals. [21,40,41] This method was

---

[d] The Favorskii reaction, a special example of dehydrohalogenation, is discussed in the following section.

first applied to cyclopropanones in a synthesis of the spiro compound *5* from the trithiane derivative *40*. [21] As Table 6 shows, the yields are usually high and the method appears to be general.

   gem-Dihalocyclopropanes which are readily obtained by the addition of dihalocarbenes to olefins [42-46] provide a useful route to cyclopropanones (Scheme 6). Silver ion-assisted solvolysis of *41* (X=Y=Br),

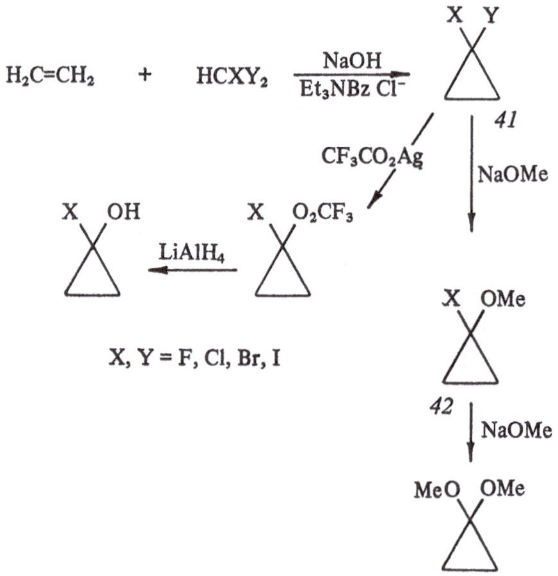

$$H_2C=CH_2 \quad + \quad HCXY_2 \quad \xrightarrow[\text{Et}_3\text{NBz Cl}^-]{\text{NaOH}} \quad$$

X, Y = F, Cl, Br, I

Scheme 6

followed by reduction, as shown, leads to α-bromocyclopropanol.[47] Treatment of *41* with excess methoxide or methyl mercaptide affords the ketal or thioketal of cyclopropanone via the correponding α-haloether *42*.[45a, 46]

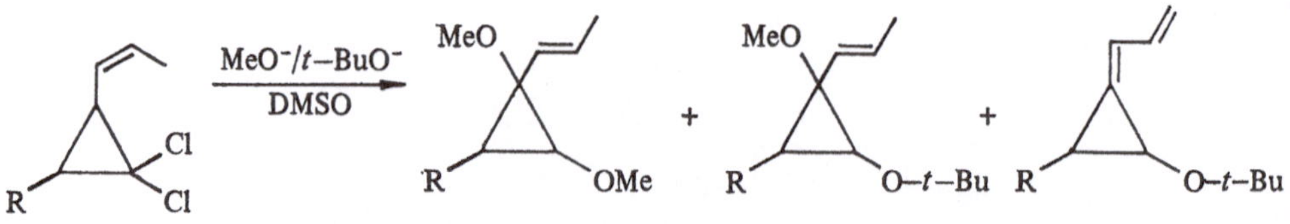

13

44   *45*   *46*

a, R = Me
b, R = H

Table 5. Cyclopropanone ethylene ketals [11]

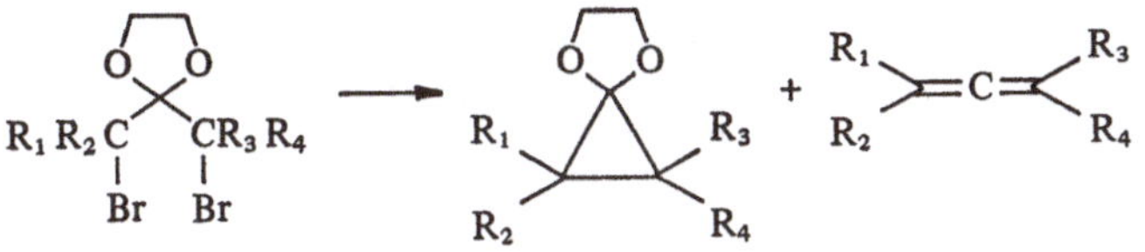

	Products (% yields)			
	Mg/THF		Zn/HMPT	
Dibromide	Allene	Ketal	Allene	Ketal
$R_1=R_2=R_3=R_4=H$	44	50	20	20
$R_1=R_2=R_3=H,R_4=Me$	47	37	20	23
$R_1=R_2=R_3=H,R_4=Et$	39	53	40	23
$R_1=R_2=H,R_3=R_4=Me$	37	40	50	28
$R_1=R_3=H,R_2=Me,R_4=Pr$	20	58	60	28
$R_1=R_2=R_3=H,R_4=C_6H_{13}$	30	60	60	20
$R_1=R_2=R_3=H,R_4=Ph$	19	70	—	—
$R_1=R_3=H,R_2=R_4=Me$	—	70	—	—
$R_1=R_3=H,R_2=R_4=Et$	10	75	70	20
$R_1=R_3=H,R_2-R_4=(CH_2)_3$	—	93	—	—
$R_1=R_3=H,R_2-R_4=(CH_2)_4$	—	83	—	—

On the other hand, the dichloro compounds *43a* and *b* are converted to the 1,2-dialkoxycyclopropanones *44* and *45* and the alkoxydiene *46*.[48] These displacement reactions appear to involve dehydrohalogenation to the cyclopropene, double bond migration and addition as illustrated in the case of *47* (Scheme 7).

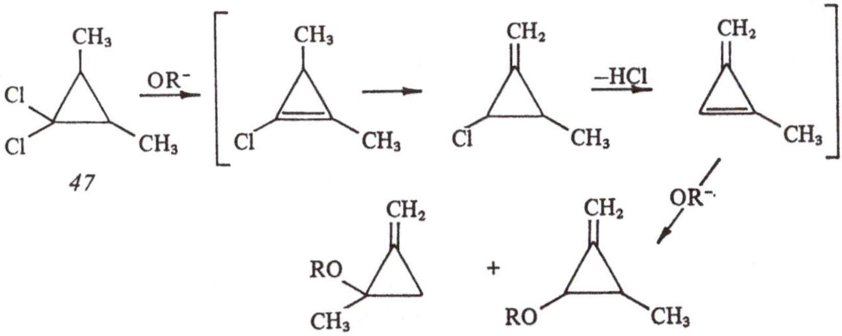

Scheme 7

Table 6. Cyclopropanone dithioketals from base catalyzed cyclization of 3-halo-alkyl dithioketals

Halo compound	Base	Cyclopropanone dithioketal	Yield, %	Ref.
	KNH$_2$		75%	21)
R=Me	BuLi		90	40)
R=(CH$_2$)$_3$CHMe$_2$	BuLi		80	40)
R=(CH$_2$)$_3$$\overset{\text{Cl}}{\text{C}}Me_2$	BuLi		85	40)
R=(CH$_2$)$_3$$\overset{\text{Me}}{\underset{\text{Cl}}{\text{C}}}$(CH$_2$)$_3$$\overset{\text{Cl}}{\text{C}}Me_2$	BuLi		65	40)
R$_1$,R$_2$=H,R$_3$=Me	BuLi		73	41)
R$_1$=R$_2$=H,R$_3$=Et	BuLi		71	41)
R$_1$=R$_2$=H,R$_3$=i—Pr	BuLi		60	41)
R$_1$=R$_2$=H,R$_3$=Bu	BuLi		72	41)
R$_1$=H,R$_2$=Me,R$_3$=Et	BuLi		63—68	41)
R$_1$=H,R$_2$=Et,R$_3$=Me	BuLi		71	41)
R$_1$=R$_2$=R$_3$=Me	BuLi		70	41)
R$_1$=R$_3$=Me,R$_2$=Et	BuLi		57—59	41)

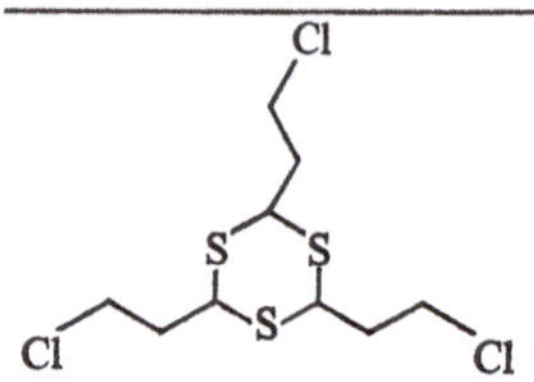

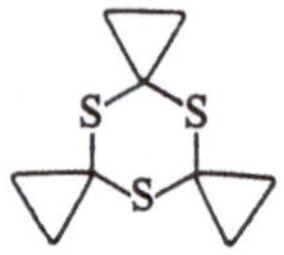

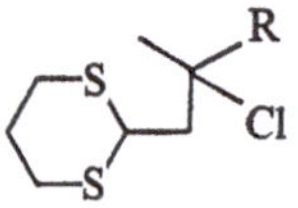

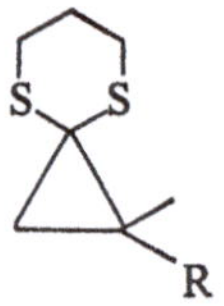

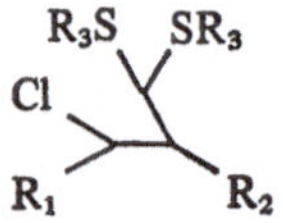

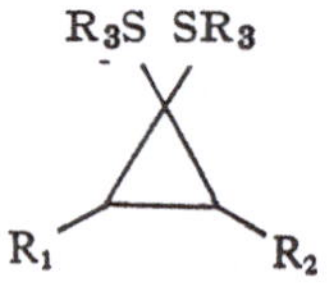

2.5. Favorskii Reactions

The classic labeling studies of Loftfield [49a)] have demonstrated that cyclopropanones are intermediates in the Favorskii reaction [49 b)] — the base-induced rearrangement of α-haloketones (Scheme 8). The related reaction of α,α'-dibromoketones has, in fact, become a convenient preparative route for cyclopropenones, e.g., 48 → 49 via 50.[50)]

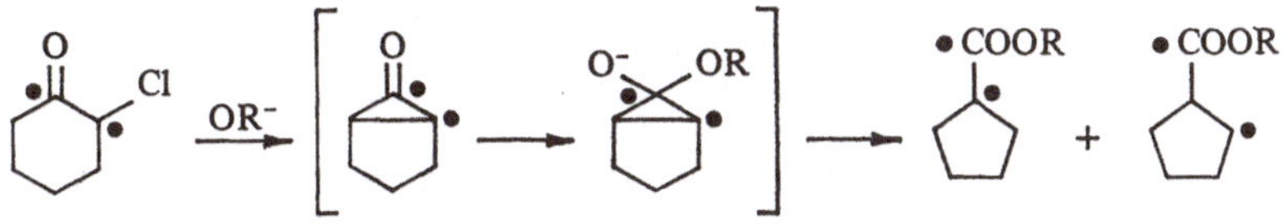

Scheme 8 · ≡ carbon—14

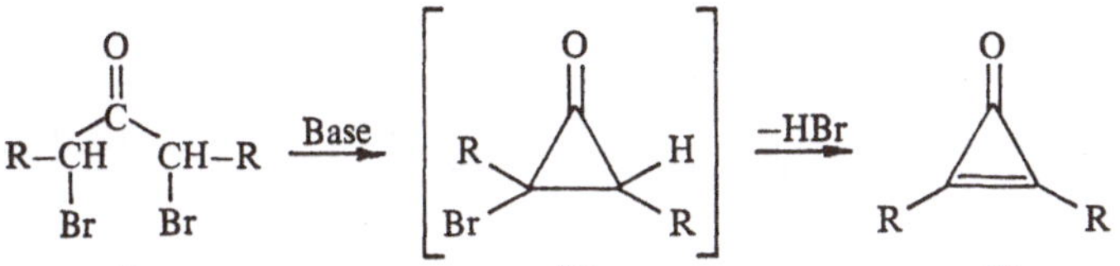

48   50   49

The first cyclopropanone to be isolated under Favorskii conditions was obtained from the reaction of the sterically hindered α-bromodi*neo*pentyl ketone (*51*) with potassium *p*-chlorophenyl-dimethylcarbinolate.[13] The product, *trans*-2,3-di-*t*-butylcyclopropanone (*52*) (20—40%) was later prepared independently by the valence isomerism of 1,3-di-*t*-butyl-allene oxide [51] (see Section 2.6).

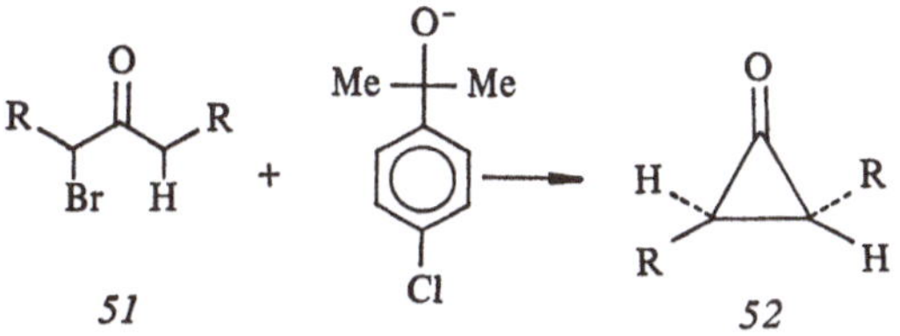

51   52

R = *t*–Bu

A Favorskii type of reaction has also been applied to the synthesis of novel fused-ring cyclopropanone derivatives. Thus, 6,6-dipiperidino-bicyclo-[3.1.0]-hexane (*53*) is obtained (34%) from α-chlorocyclohexanone and excess piperidine.[52a] The other major product (41%) is 2-piperidin-ocyclohexanone. Similarly, α-chlorocycloheptanone reacts with pyrroli-dine affording 7,7-dipyrrolidino-bicyclo[4.1.0]-heptane (*54*).[52b] The intermediate in these reactions is very probably the cyclopropyl iminium salt, *e.g.* *55*. Compounds *53* and *54* are both stable, low melting solids which may be used as *in situ* sources of the corresponding cyclopropanones (see Section 4.1.3). Although *53* reverts to α-chlorocyclohexanone with aqueous hydrochloric acid, *54* undergoes hydrolysis to 7-pyrrolidino-7-

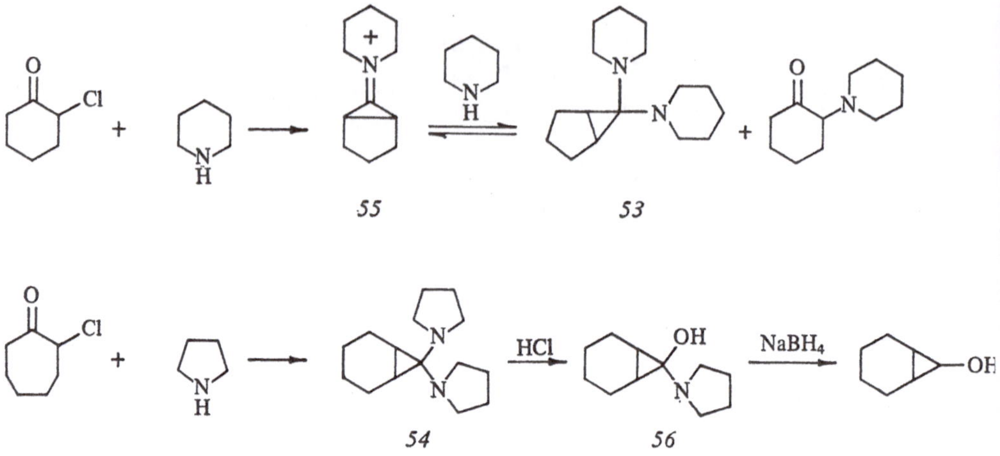

55   *53*

54   *56*

norcaranol (*56*) which in turn yields *endo*-7-norcaranol upon reduction with sodium borohydride.[52b)]

Recently, cyclopropanimines *57 a* and *b* have been prepared (93—96%) from the reaction of potassium *t*-butoxide with α-haloimines.[53)] Upon heating, *57* decomposes to *t*-butylethylene and 2,4-dimethyl-3-pentyl*iso*cyanide (*58*). The imine is also surprisingly resistant to alkaline hydrolysis in dioxane-water; conversion to the amide *59* requires heating at 100 °C for 24 h (Scheme 9).

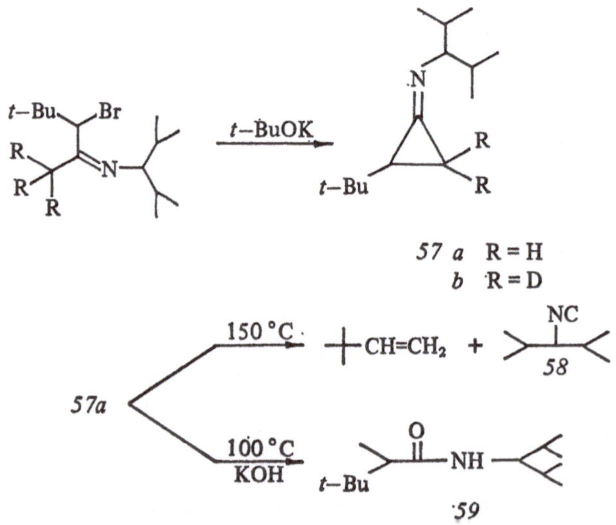

57 a   R = H
   b   R = D

Scheme 9

2.6. Miscellaneous Methods

2.6.1. Valence Isomerism of Allene Oxides and Related Systems

Another possible route to cyclopropanones is via the isomerism of allene oxides, *i.e.*, *60* ⇌ *61*. While valence isomerism has been established in related systems such as *62* ⇌ *63* [54], only two cyclopropanones have as yet been prepared by this method and its usefulness appears limited.

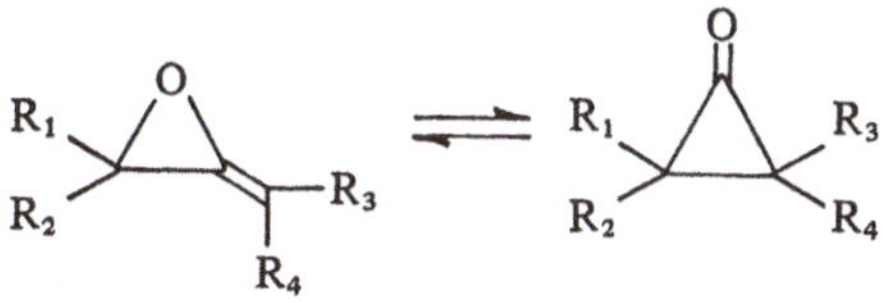

60 *61*

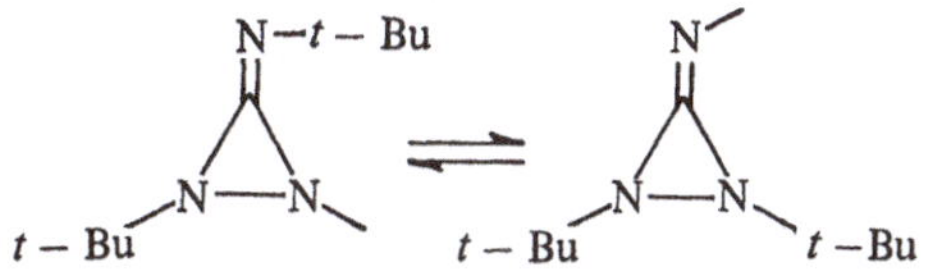

62 *63*

1,3-Di-*t*-butylallene oxide (*64*) is an isolable compound prepared by the oxidation of 1,3-di-*t*-butylallene with *m*-chloroperbenzoic acid. Upon heating at 100 °C for 5 h, 50% of *64* is isomerized to *trans*-2,3-di-*t*-butylcyclopropanone (*52*).[51] Likewise, oxidation of 1,1-di-*t*-butyl-allene (*65*) with buffered peracetic acid (equimolar quantities) affords 2,2-di-*t*-butylcyclopropanone (*66*).[55a] Although the intermediate is presumably 3,3-di-*t*-butylallene oxide, it has not been detected.

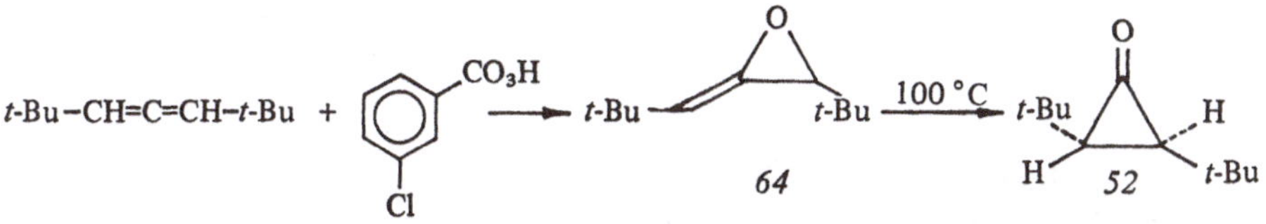

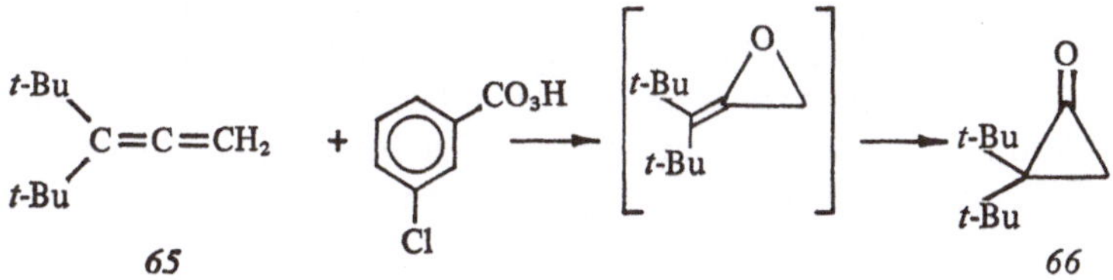

64 *52*

65 *66*

On the other hand, treatment of the mono-*t*-butylallene oxide *67* with one equivalent of peracid yields the spiro-*bis*-epoxide *68* and the acetoxy ketone *69*.[55b] Similarly, the peracid oxidations of tetramethyl-allene, 1,1-dimethylallene, and 1,2-cyclononadiene do not give the

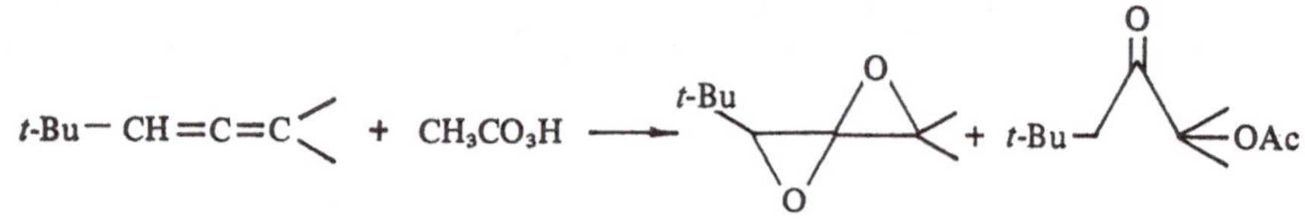

$$t\text{-Bu}-\text{CH}=\text{C}=\text{C} \diagdown \quad + \quad \text{CH}_3\text{CO}_3\text{H} \longrightarrow$$

67 *68* *69*

corresponding cyclopropanones but a mixture of products [55c] consistent with the reaction sequence shown in Scheme 10. The intermediate allene oxide *70* appears to decompose by two routes: one, *via* the cyclopropanone *71*, the other, *via* the dioxaspiropentane *72*. The cyclopropanone is then converted to products by oxidative decarbonylation, ring opening, and a Baeyer-Villiger reaction. The oxide *70* has been prepared in solution

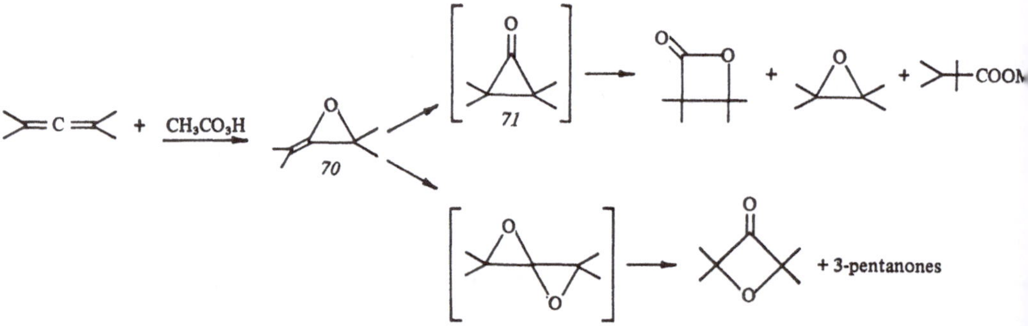

$$\text{>C}=\text{C< } + \text{ CH}_3\text{CO}_3\text{H} \longrightarrow$$

70 *71* *72*

Scheme 10

by another method and its mass spectrum provides additional evidence for the rearrangement to *71*.[56]

In connection with the above discussion, formation of 3,3-disubstituted 2 (3 H)-oxepinones (*73*) in the dye-sensitized photooxygenation of 6,6-disubstituted fulvenes is of special interest. [57,58] The reaction may be pictured in terms of an allene oxide intermediate which, as shown in Scheme 11, isomerizes to a cyclopropanone, followed by intramolecular rearrangement.

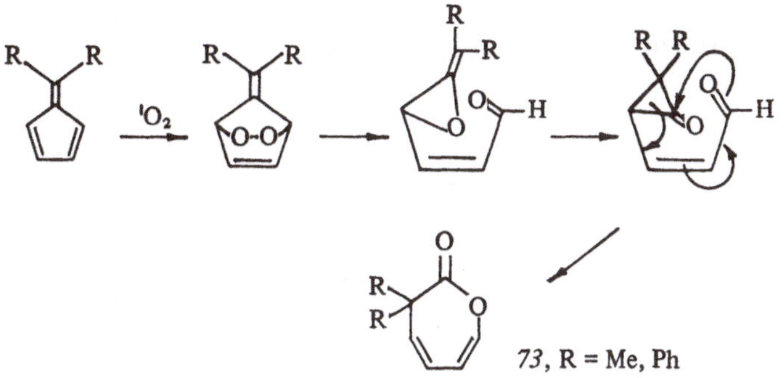

73, R = Me, Ph

Scheme 11

There has been interest in the possibility of an equilibrium between unsaturated aziridines and cyclopropanimines analogous to the allene-oxide-cyclopropanone rearrangement [53,59,60], and evidence for such an interconversion has recently been obtained.[60] Thus, like the cyclopropanimine 57 (Scheme 9), the aziridines 74a—d undergo thermal decomposition to olefins and isocyanides. The rearrangement of 74 in this manner strongly suggests the intermediacy of the cyclopropanimine 75. In fact, compounds corresponding to 75b—d have been characterized by NMR upon heating solutions of 74. The reverse reaction, cyclopropanimine to aziridine, has as yet not been observed.[53,60]

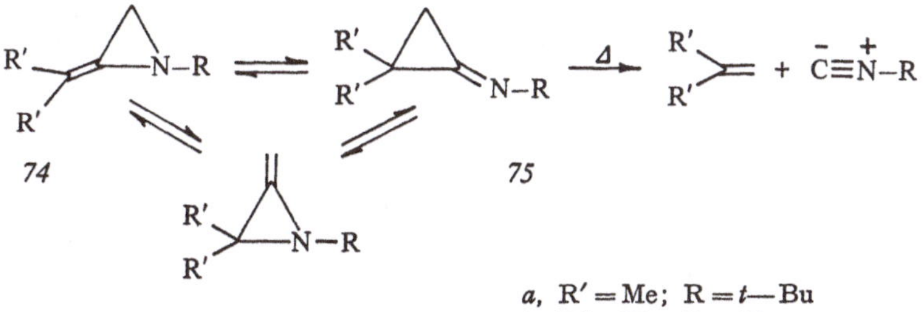

74   75

a, R′ = Me; R = t—Bu
b, R′ = H; R = Me
c, R′ = H; R = Et
d, R′ = H; R = t—Bu

2.6.2. Cyclopropanones from Cyclopropenones

3,3-Dimethoxycyclopropene has been converted into a number of cyclopropanone ketals as shown in Scheme 12.[61] Alternatively, the ketal 76 may be directly obtained from the cyclopropene precursor *77* by treatment with potassium *t*-butoxide.

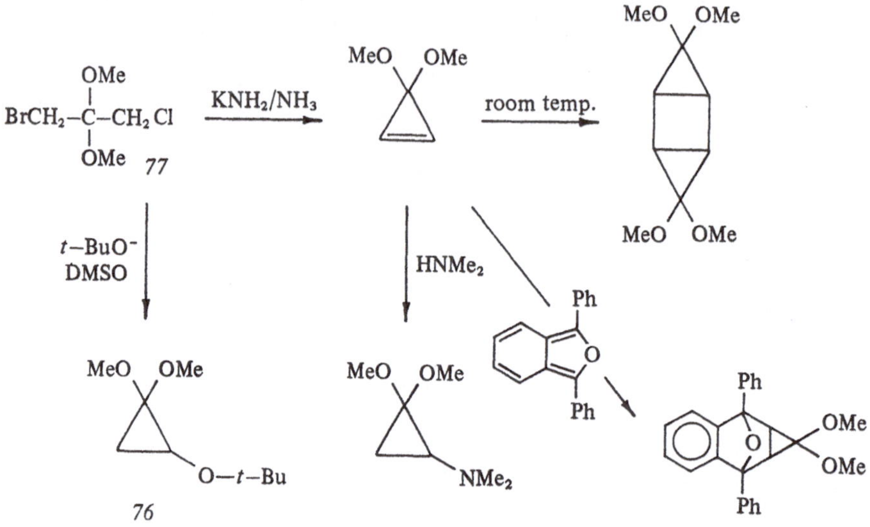

Scheme 12

Recently, the bicyclic cyclopropanones *78* and *79* were prepared in high yield by the Diels-Alder addition of cyclopropenone to the diphenylisobenzofuran *80* and the dimethylanthracene *81*.[62]

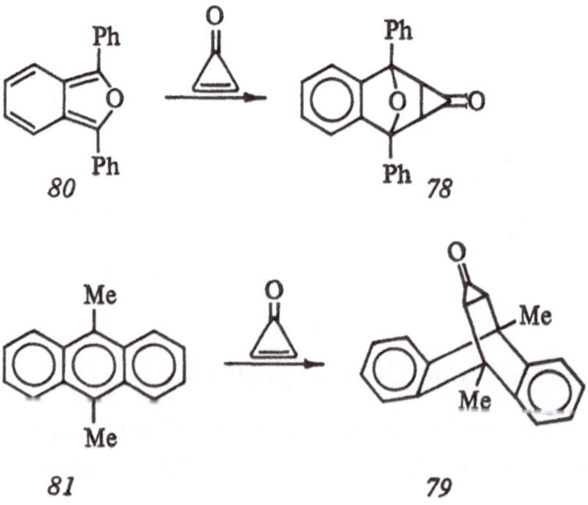

3. Spectroscopic Properties of Cyclopropanones

Since most cyclopropanones are too unstable to permit isolation and purification, structural aids such as X-ray diffraction or even routine elemental analyses may not generally be employed for identification. Consequently, investigators have usually relied on spectroscopic characterization in solution. For this purpose, infrared and microwave spectroscopy have proved particularly valuable. The infrared spectra of cyclopropanones are distinguished by an unusually high frequency carbonyl stretching vibration (~ 1825 cm^{-1}) which provides a rapid and simple method of detection.[10] Complimenting this, microwave spectroscopy has enabled a precise determination of such structural parameters as bond lengths and bond angles.[63] Other forms of spectroscopy — such as ORD [64], UV [32], and NMR [10] — have also been used for structural and chemical studies.

3.1. Microwave Spectra

Pochan, Baldwin and Flygare have analyzed the microwave spectra of cyclopropanone and the isotopic isomers $^{13}C_1$, $^{13}C_2$, and 2,2-dideutero-cyclopropanone.[63] The rotational transitions were determined by studying the Stark effect (the shifts and splittings of lines produced by an electric field). The type of transition observed for cyclopropanone was consistent with C_{2v} symmetry and the sum of the moments of inertia $(I_a + I_b - I_c)$ suggested that all four protons are out-of-plane. These data eliminate such structural alternatives as the dipolar oxyallyl tautomer *82* and allene oxide *83*. The electric dipole moment (μ_a) was calculated to be 2.67 ± 0.10 D, which corresponds to an average of those of acetone (2.93 D) [65] and formaldehyde (2.34 D).[66]

82   *83*

Further analysis of the microwave spectra provides a precise description of the bond lengths and angles (Table 7). The C_2C_3 bond is unusually long and relatively weak as shown by the reactivity of substituted cyclopropanones in cycloaddition reactions (see Section 4.4). The carbon-oxygen bond is somewhat shorter than the average carbonyl group and this feature is reflected in the infrared properties of cyclopropanones outlined below.

101

H. H. Wasserman, G. M. Clark, and P. C. Turley

3.2. Infrared Spectra

A characteristic feature of the infrared spectra of cyclopropanones is the unusual position of the carbonyl stretching vibration. Whereas simple ketones absorb in the region 1725—1705 cm^{-1} [67)], three-membered cyclic ketones exhibit bands in the 1875—1796 cm^{-1} range (Table 8).

Table 7. Bond distances and bond angles of cyclopropanones

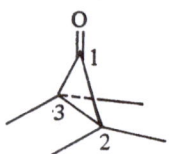

	Cyclopropanone[63)]	Model[1)]
Bond distances (Å)		
C_1—O	1.191 ± 0.020 Å	1.23 ± 0.03, acetone
C_1—C_2	1.475 ± 0.017 Å	1.52 ± 0.03, acetone
C_2—C_3	1.575 ± 0.012 Å	1.53 ± 0.01, cyclopropane
C_2—H	1.086 ± 0.025 Å	1.09 (assumed) acetone
Bond angles		
$C_1C_2C_3$	57° 42′ ± 25′	
$C_2C_1C_3$	64° 36′ ± 50′	93.1 ± 0.3°, cyclobutanone[2)]
HCH plane angle with C_2—C_3 axis	29° 7′ ± 2°	
HCH	114° 8′ ± 2°	

[1)] Tables of interatomic distances and configuration in molecules and ions. London: Chemical Society 1958.
[2)] Scharpen, L. H., Laurie, V. W.: J. Chem. Phys. *49*, 221 (1968).

   This band often appears as a doublet probably resulting from the accidental degeneracy of an overtone of another low-lying vibration mode (Fermi resonance).[67a, 68)] The C—H stretching vibrations also undergo a high-frequency shift and appear at ~ 3050 cm^{-1}, a position generally characteristic of the methylene stretch in cyclopropyl systems.[67)]
   The high-frequency shifts in the CO and CH stretching vibrations of cyclopropanones are related to ring strain and occur to a lesser extent in cyclobutanone. Although normal coordinate analysis has been employed in predicting these shifts, the theoretical models do not appear to be completely reliable.[69)] An important factor causing the above shifts

appears to be a strain-induced rehydridization of the ring bonds accompanying a decrease in ring size. This leads to an increase in the p-character of the ring bonds and, consequently, an increase in the s-character of the exocyclic bonds resulting in an increase in the CO and CH bond strengths.

Table 8. IR bands of cyclopropanones

Compound	Bands, (cm^{-1})		Ref.
	CH$_2$Cl$_2$	3045 (C—H)	[2,10]
		1813 (C=O)	
	CCl$_4$	1816 (C=O)	[3]
	CHCl$_3$	1908, 1822 (C=O)	[68]
	CH$_2$Cl$_2$	1850, 1822 (C=O)	[10]
	CH$_2$Cl$_2$	3050 (C—H)	[10]
		1815 (C=O)	
		1387—1380 (d, C(CH$_3$)$_2$ bending)	
	CH$_2$Cl$_2$	1843, 1823 (C=O)	[10]
		1822 (C=O)	[96]
		1387, 1380 (C(CH$_3$)$_2$ bending)	
		1825 (C=O)	[55a]
	CCl$_4$	1822 (C=O)	[13]

3.3. Electronic Absorbtion Spectra (UV and ORD)

The n → π* transition in cyclopropanone occurs at 310 nm ($\varepsilon = 23$) (compared to 280 nm ($\varepsilon = 15$) for simple ketones) [70] and undergoes a shift to longer wavelengths with successive alkylation of the α-carbons (Table 9). Turro has attributed the longer wavelength of this transition

Table 9. n→π* transitions in cyclopropanones

Compound	Solvent	λ_{max}, nm	ε, M^{-1} cm^{-1}	Ref.
	CH_2Cl_2	310 330 (sh)	23 15	[2]
	CH_2Cl_2	330	18	[10]
	CH_2Cl_2	340	27	[10]
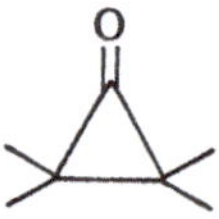	CH_2Cl_2	340 (350)	20	[10]
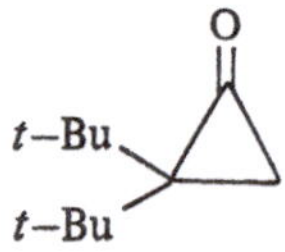	Hexane	345	52	[55a]
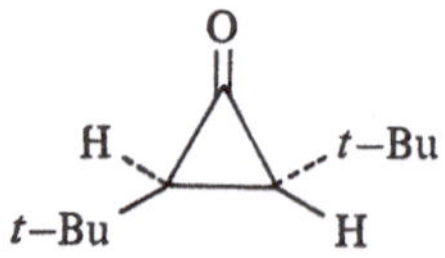	Isooctane	354	33	[13]

in cyclopropanones to a strain-induced destabilization of the ground state and a favorable charge situation in the excited state.[32] However, on examination of a series of cycloalkanones one does not observe correlation of shifts in the n → π* absorption with ring size.[71]

Optically active *trans*-2,3-di-*t*-butylcyclopropanone has been prepared by asymmetric destruction of the racemic compound with *d*-amphetamine.[64] The (+)-cyclopropanone, $[\alpha]_{436}^{25} + 76°$ (*c* 0.5, CCl$_4$) exhibits a positive Cotton effect with a peak at 370 nm. Racemization occurs upon heating.

3.4. Nuclear Magnetic Resonance Spectra

The protons of cyclopropanones show absorption shifted upfield from those in unstrained ketones and display signals in the range 1.1—2.1 ppm (Table 10). This is not unexpected since a shielding field is associated with the cyclopropane ring.[72] Carbon-13 coupling constants have also been measured for the ring protons of 2,2-dimethylcyclopropanone. The splitting ($J(^{13}CH) = 160$ Hz)[10] is the same as that found in cyclopropane (161 Hz).[73]

A dynamic equilibrium between cyclopropanone and certain hemiacetal and related derivatives may be observed by NMR. While the ring protons of 1-acetoxycyclopropanol exhibit an A$_2$B$_2$ pattern indicating a slow interconversion with the ketone, they appear as a sharp singlet in the methyl and ethyl hemiacetals in accord with a rapid equilibration with cyclopropanone.[4,10] Other cases indicate that the interconversion process is base-inhibited [5] and requires, as substituents at the 1-position, a good leaving group and a heteroatom capable of multiple bonding (see Section 4.1.3).[6,74]

4. Reactions of Cyclopropanones

As highly reactive systems, cyclopropanones undergo a number of varied reactions. For example, thermal decomposition of *3* may proceed by three distinct pathways — decarbonylation [62,64], ring opening [1,2,34], or in the absence of stabilizers, polymerization. [3,10] Similarly, there are two modes of photochemical decomposition — decarbonylation and polymerization. [10,75] The carbonyl group of *3* is subject to attack by several different types of nucleophiles and many of the 1-substituted cyclopropanols undergo ready rearrangement accompanied by ring enlargement. [76]

Ring opening reactions constitute a major part of cyclopropanone chemistry. All of the ring bonds are labile, but cleavage is generally selective for a given reaction. A C$_1$—C$_2$ ring opening usually requires an initial attack on the carbonyl group as exemplified by the second stage of the Favorskii rearrangement (Scheme 7). On the other hand, many reactions of *3* involving C$_2$—C$_3$ cleavage are best pictured in terms of the oxyallyl tautomer. Theoretical calculations suggest that 78 kcal/mol are required for isomerization in the parent system [77], this value

Table 10. NMR data for cyclopropanones

Compound	Solvent	δ ppm (TMS)				Ref.
		H_a	H_b	H_c	H_d	
	CH$_2$Cl$_2$ Neat or CFCl$_3$	1.65(s) 1.72(s)				10) 3)
	CH$_2$Cl$_2$	$\begin{cases}1.85{-}2.1 \text{ (m, 1 H)} \\ 0.9{-}1.8 \text{ (m, 4 H)}\end{cases}$				10)
	CH$_2$Cl$_2$	1.20(s)	1.40(s)			10)
	CH$_2$Cl$_2$ ^{1})	1.1– 1.4(m)	1.22(d, $J_{ab}=5$)	1.12(s)	1.36(s)	96)
	CH$_2$Cl$_2$		1.19(s)			10)
		1.44(s)	1.10(s)			55a)
	CCl$_4$	1.55(s)	0.96(s)			13)
		2.75(s, J ^{13}CH = 173; J_{HH} = 9.0)				62)

^{1}) c,d assignments are arbitrary and may be reversed.

decreasing markedly with successive alkyl substitution of the C_2 and C_3 carbon atoms. [78] (In the case of *trans*-2,3-di-*t*-butylcyclopropanone, experimental evidence indicates that this process requires a free energy of activation equal to 27 kcal/mol. [64]) These predictions are consistent with the relative reactivity of cyclopropanones toward $4+3 \rightarrow 7$ cyclo-additions which involve oxyallylic character in the transition state and possibly an oxyallylic intermediate. [79]

4.1. Reactions of the Carbonyl Group

The carbonyl group in cyclopropanone readily adds many types of nucleo-philes, even at low temperature, *e.g.* water, amines, acids, Grignard reagents. The unusual tendency toward adduct formation extends to polymeriza-tion and is a consequence of the strain energy released by the $sp^2 \rightarrow sp^3$ rehybridization of a carbon atom constrained by a three-membered ring.

4.1.1. Polymerization

In the absence of inhibitors, polycyclopropanone (*84*) is readily formed when the monomeric solution is warmed above 0 °C. [1-3] [e]
The polymer is a white solid of varying composition (reported mole-cular weights: 3000—4000 [10], ~9500 [3]) which is soluble in benzene, chloroform, and methylene chloride, but insoluble in ether and acetone. The spectroscopic characteristics of the polymer which contains ~170 cyclopropanone units include a broadened NMR singlet (CDCl$_3$) at 1 ppm and IR bands (CHCl$_3$) at 1450, 1310, 1130 (C—O stretch), 1010, 975 and 950 cm^{-1} (C—C stretch). [3]

The polymerization of cyclopropanone is initiated [80] by traces of water and is inhibited [15] by moisture scavengers such as acetyl chloride. The terminal groups are apparently hemiketal units since α,ω-diacetoxy-poly(oxycyclopropylidenes) *85* are isolated from a mixture of cyclopro-panone hydrate, diazomethane and excess ketene. [80]

84

---

[e] Polycyclopropanone resulting from ^{60}Co irradiation of a carbon monoxide-ethylene mixture has also been reported (Murahashi, S., Takizawa, T., Asada, T.: Japanese Patent 17636 (1961); Chem. Abst. *56*, 15679h (1962)).

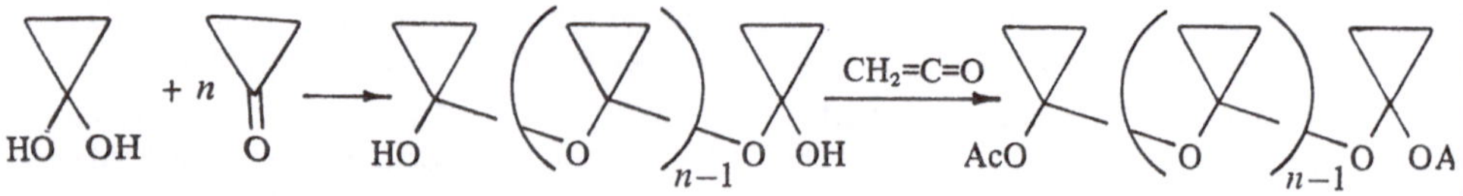

4.1.2. Reactions with Water, Alcohols and Acids

The hydrate and hemiacetals of cyclopropanone are rapidly formed in good yields by the addition of water and alcohol, respectively, to a solution of the ketone. [1-3,10] These derivatives may be isolated and purified, but they tend to rearrange to propionic acids or esters when stored at room temperature (Scheme 13).

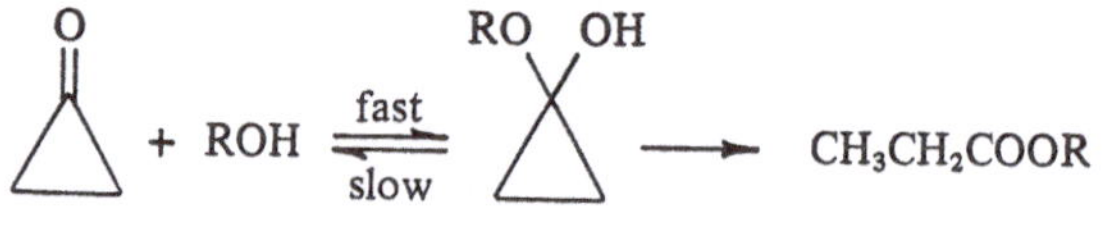

Scheme 13

Since the hemiketals of cyclopropanone are in equilibrium with the parent ketone, they may be interconverted by standing in the presence of the appropriate alcohol. [4] However, in the cases of cyclopropanone ethyl and methyl hemiketals, dissociation to the ketone is slow and hemiketal interconversion requires about two weeks at room temperature for completion. [4]

Other hemiacetals which have been prepared include the methyl [8,9a], ethyl [9b], and isopropyl hemiacetals [81] of tetramethylcyclopropanone, the phenyl, α- and β-naphthyl hemiacetals of cyclopropanone [82], and the benzyl hemiacetal of 2,3-di-t-butylcyclopropanone. [13] In the last case, the benzylic methylene protons display an AB pattern in the PMR spectrum indicating that the two t-butyl groups are *trans* to one another. Although derivatization of this di-t-butyl ketone was possible, carbonyl addition may be hindered by steric factors as suggested by the lack of reaction of 2,2-di-t-butylcyclopropanone with methanol. [55a]

Cyclopropanone hemiacetals and hydrate react with ketene to form acetoxy derivatives, e.g. 86 and 87, respectively. [5,15,80] As shown in Scheme 14, the diacetoxy compound 87 may also be obtained from the reaction of cyclopropanone with acetic acid and excess ketene. [83]

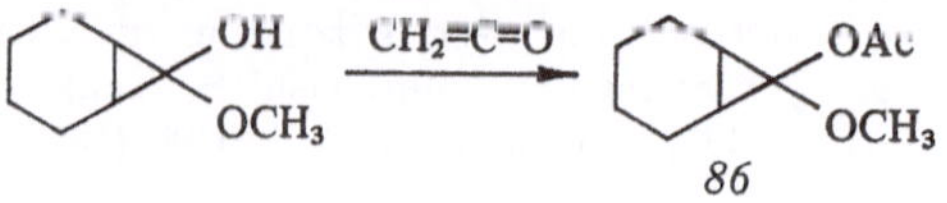

Presumably, 1-acetoxycyclopropanol (*88*) is the intermediate in this reaction as well as that between the hydrate and ketene. [80] Although *88* has not been isolated in the above cases, it may be prepared from acetic acid and cyclopropanone and reacts with ketene to give *87*. [5,15]

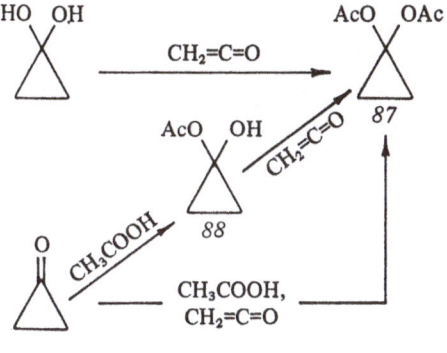

Scheme 14

Cyclopropanone forms a carbonyl adduct with hydrogen cyanide [83] which hydrolyzes readily in acid to give the surprisingly stable hydroxy acid *89*. [84] Attempts to form the acylium ion of *89* by reaction with concentrated sulfuric acid or other activating agent such as dicyclohexylcarbodiimide were unsuccessful. In fact, not only is *89* unreactive toward ring enlargement, but, even under neutral conditions in aqueous solution, 1,2-cyclobutadione undergoes a benzilic acid type rearrangement to form *89*. [85]

The reaction of cyclopropanone with hydrogen chloride affords the chlorohydrin *90* (65%). [5,15] Acetylation of *90* with ketene gives the expected acetoxy compound *91* (34%) while the addition of methanol to a methylene chloride solution of *90* affords cyclopropanone methyl hemiketal (100%). The latter reaction illustrates the use of *91* as an *in situ* source of cyclopropanone.

In contrast to the above, 2,2-dimethylcyclopropanone yields the ring opened products *92* and *93* with hydrochloric acid and acetic acids rather than the cyclopropyl adducts. [5] This preference for cleavage is consistent with the enhanced stability of the dimethylhydroxyallyl cation compared to the unsubstituted ion.

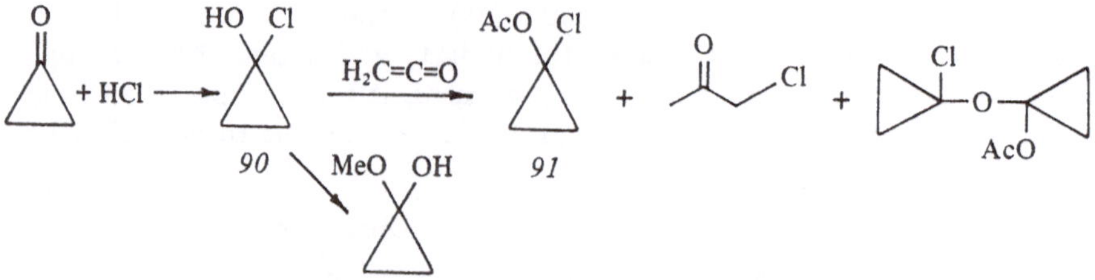

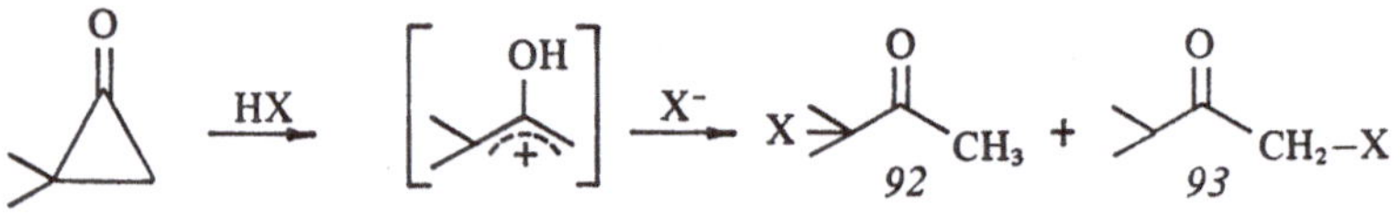

4.1.3. Reactions with Amines, Cyclopropanimines

Cyclopropanones react with ammonia and aliphatic amines to form carbinol amines which usually undergo further amine substitution. For example, attempts to prepare the carbinol amines [f] from dimethylamine and cyclopropanone resulted only in the isolation of the 1,1-diamino derivative 94. [86] When methyl amine is added to cyclopropanone, the

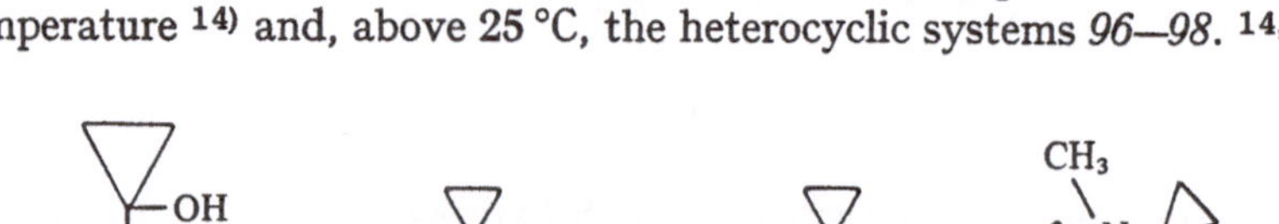

carbinol amine reacts further to form the *bis* addition product 95 at low temperature [14] and, above 25 °C, the heterocyclic systems 96–98. [14,86]

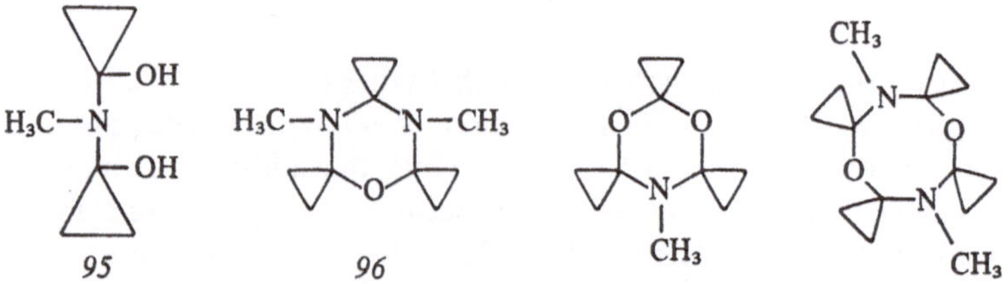

---

[f] 1-Dimethylamino-1-hydroxycyclopropane has been prepared from 1-acetoxy-cyclopropanol.[14]

Similarly, the reaction with ammonia affords di- and tri-1-hydroxycyclo-propyl amines at $-30\,°C$ while at $30\,°C$ ring opening to propionamide occurs (Scheme 15). [14]

On the other hand, some carbinol amines are sufficiently stable to permit spectroscopic characterization, *e.g.*, 1-cyclohexylamino- and

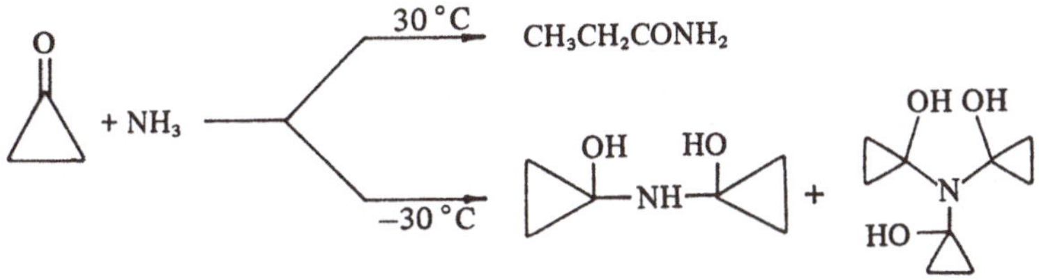

Scheme 15

1-*t*-butylaminocyclopropanols [87], while others may be isolated, *e.g.*, 1-piperidino- and 1-morpholinocyclopropanols. [6]

There is much evidence to suggest that a cyclopropanimine inter-mediate, *e.g.*, *99*, is involved in the transformation of the carbinol amines to diamines:

a) Cyclopropyl rings are extremely resistant to S_N2 reactions [88] and nucleophilic substitution is more likely to occur by a dissociative mechan-ism (*e.g.*, Scheme 16);

b) N,N-dimethylcyclopropanium ion (*99*), prepared by partial deamina-tion of *94*, has been characterized by NMR and readily adds methoxide ion to form 1-methoxy-1-dimethylaminocyclopropane [74];

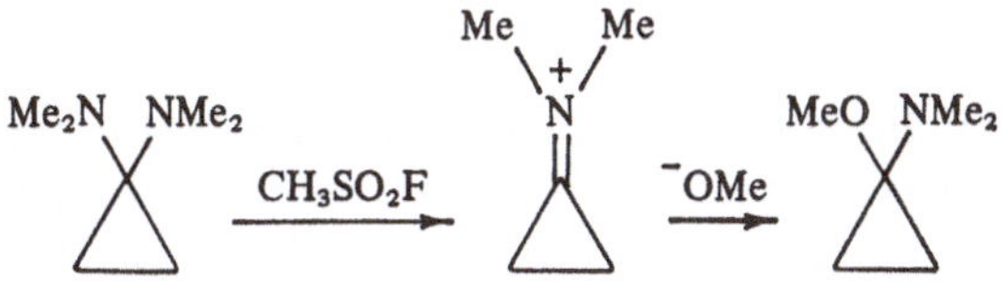

Scheme 16

c) in dilute acid, the NMR behavior of 1-piperidinocyclopropanol (*100*) indicates the existence of a dynamic equilibrium between carbinol amine and imine which is retarded in nonpolar solvents. [6] When *100* is dissolved in a piperidine solution, it further reacts to form the *bis*-piperidino adduct *101*.

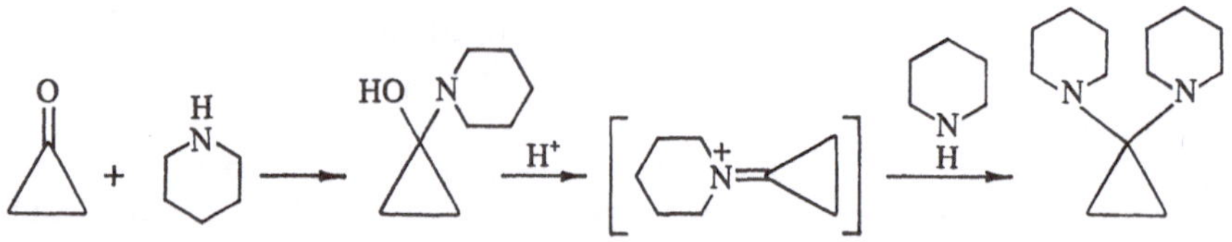

In some cases, 1,1-diaminocyclopropanes may be converted to the corresponding carbinol amines. Thus, as illustrated in Scheme 17, the 7-hydroxy-7-pyrrolidinobicycloheptane *56* [52b)] is obtained from the diamino compound *54* in aqueous acid, and reacts with methanol or ethanol affording the corresponding amino ethers *102a,b*. [89)] In the presence of methyl iodide, deamination occurs and the methylhemiketal *103* is obtained (95%).

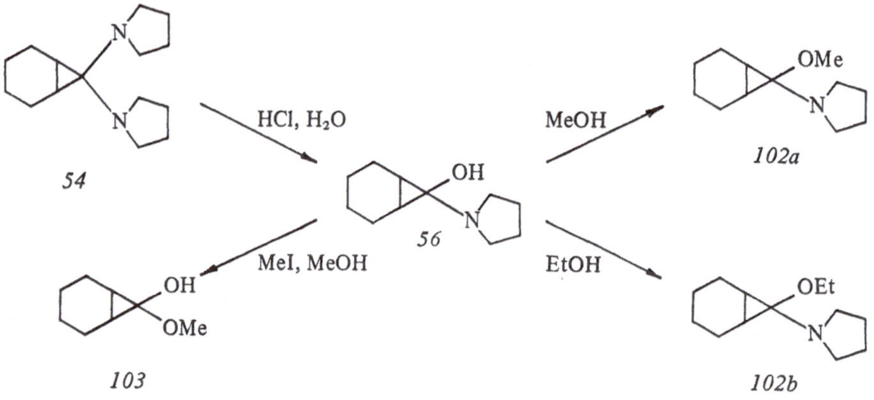

Scheme 17

N-Methylaniline adds to cyclopropanone to give the expected carbinol amine in quantitative yield. [5,90)] On the other hand, a mixture of addition products may be isolated from the reaction of cyclopropanone [5,90)] or cyclopropanone ethyl hemiketal [4)] with aniline (Scheme 18). At −78°C, the carbinol amine *104* (66%) and the dimeric species *105* (33%) were obtained when the reactants were present in stoichiometric amounts. With excess aniline, *104* was produced almost quantitatively. However, at 25 °C, both *104* and the dianilino product *106* were obtained. These products could be interconverted by the addition of aniline or cyclopropanone, respectively. At 125 °C, o propionylaminopropiophenone (*107*) was formed, possibly by further reaction of the carbinol amine with cyclopropanone as shown. [4)]

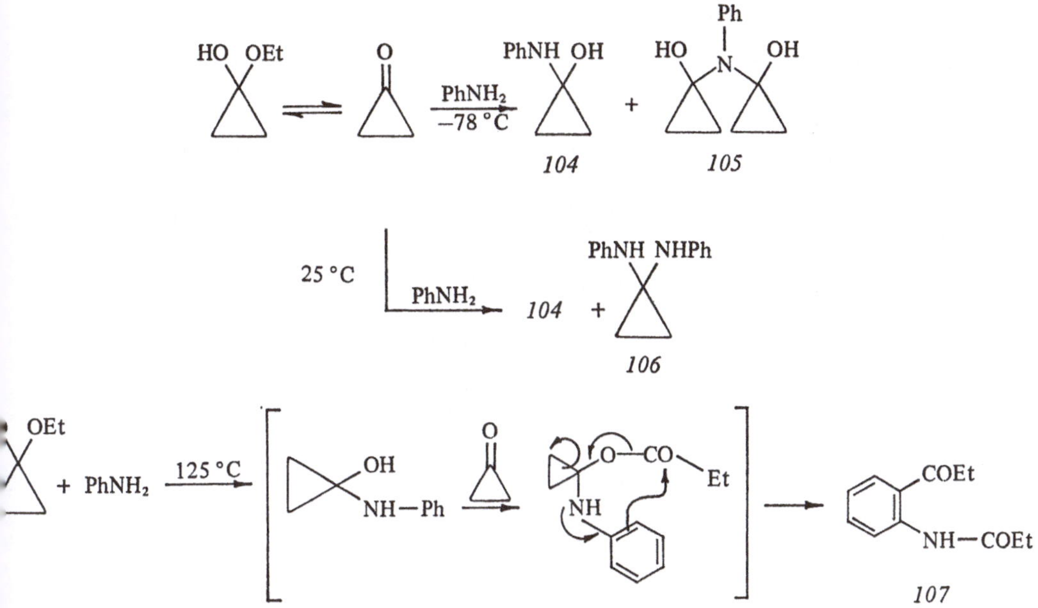

Scheme 18

4.1.4. Alkylation by Carbonyl Addition

Cyclopropanone derivatives such as 1-ethoxycyclopropyl acetate *(7 a)* and 1-ethoxycyclopropanol readily undergo addition reactions with Grignard reagents yielding a variety of 1-substituted cyclopropanols (Table 11). [4,7] With both of these cyclopropanone precursors, two moles of Grignard reagent are required, since one mole is needed to generate the intermediate free cyclopropanone (Scheme 19).

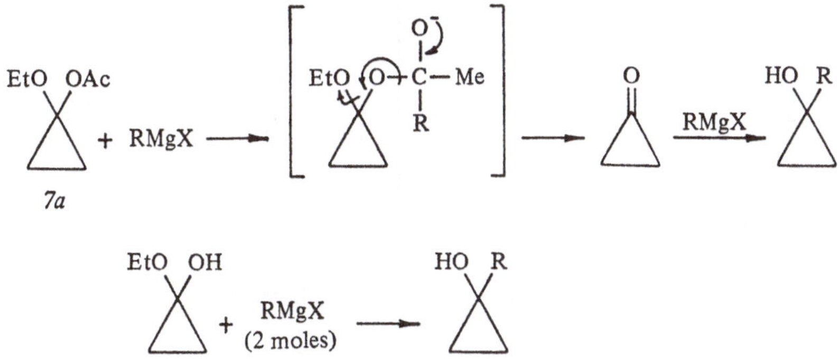

Scheme 19

Table 11. 1-Substituted cyclopropanols from Grignard addition to cyclopropanone

Cyclopropanol	Cyclopropanone precursor	Grignard reagent	Yield %	Ref.
 Me OH (cyclopropane)	 EtO OCOMe (cyclopropane)	MeMgI	46	7)
 Ph OH (cyclopropane)	,,	PhMgBr	64	7)
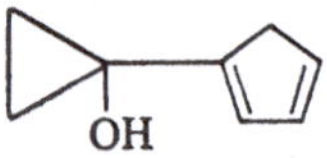 (cyclopropane)—(cyclopentadiene), OH	,,	 (cyclopentadiene)—MgBr	20	7)
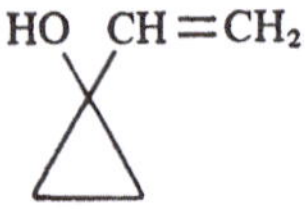 HO CH=CH₂ (cyclopropane)	 EtO OH (cyclopropane)	$H_2C=CHMgBr$	64	78)
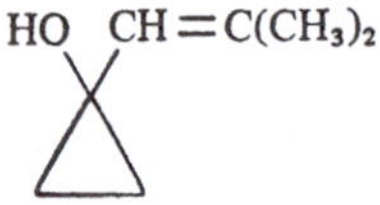 HO CH=C(CH₃)₂ (cyclopropane)	,,	$(CH_3)_2C=CHMgBr$	65	78)
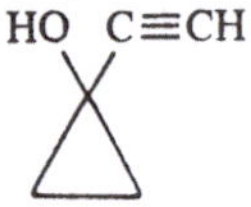 HO C≡CH (cyclopropane)	,,	$HC\equiv CMgBr$	21	78)
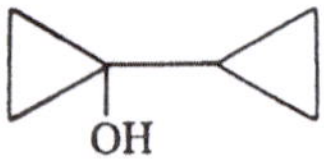 (cyclopropane)—(cyclopropane), OH	,,	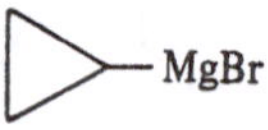 (cyclopropane)—MgBr	45	91,130)
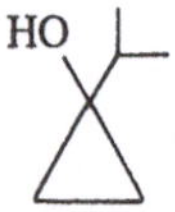 HO (cyclopropane with gem-dimethyl)	,,	$(CH_3)_2CHMgBr$		130)

Alkylation reactions which introduce a double bond adjacent to the cyclopropane ring provide intermediates which may undergo useful rearrangements to the cyclobutanone system. As shown in Scheme 20, reactions of 1-vinylcyclopropanol (*108*) with acid, positive halogens, peracids or carbonium ions lead to cyclobutanones *via* the cyclopropyl carbinyl cation *109*. [76)]

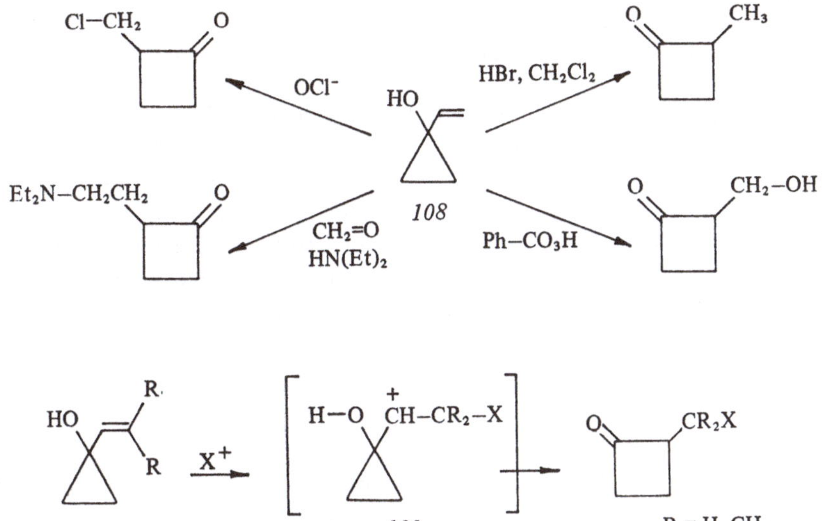

Scheme 20

In like manner, the addition product (*110*) of cyclopropanone and cyclopentadienyl Grignard reagent readily rearranges to the spiro ketone *111*. [4)]

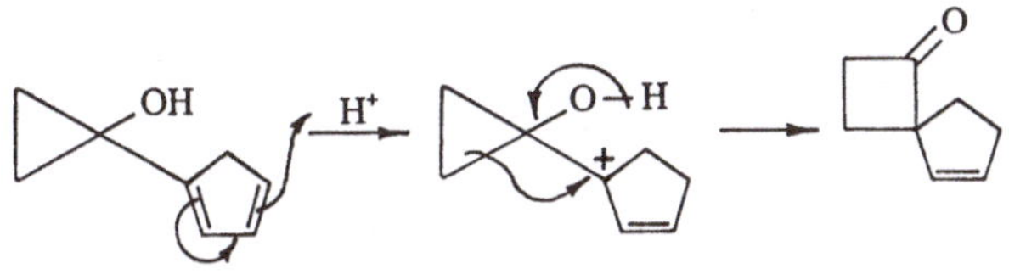

On the other hand, solvolysis of the tosyl derivative of 1-cyclopropyl-cyclopropanol (*112*) in aqueous ethanol does *not* lead to ring enlarged products. [91] Apparently, the incipient cyclopropyl cation is better stabilized by conjugative interactions with another cyclopropyl system than with a vinyl group.

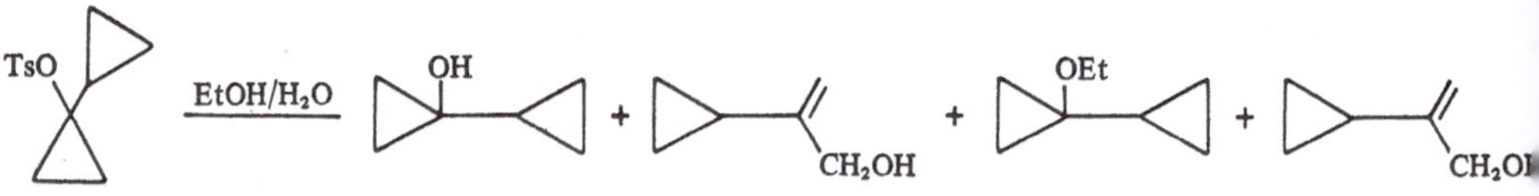

112

Other alkylation reactions are observed in the condensation of cyclo-propanium ions (generated *in situ*) with ketones [89,92], enamines [6], nitroalkanes [93], dimethylmalonate [92], and phenol. [92] Thus, 7-hydroxy-7-pyrrolidinobicyclo[4.1.0]heptane (*56*) as well as the 7,7-dipyrrolidino derivative (*54*) react with acetone to give the amino ketone *113*. [89] This reaction may be pictured as an addition of the enol form of the ketone to the reactive iminium salt formed from the carbinol amine. In like manner, phenol undergoes *ortho* substitution with the carbinol amine *114* formed from cyclopropanone and dimethyl amine.

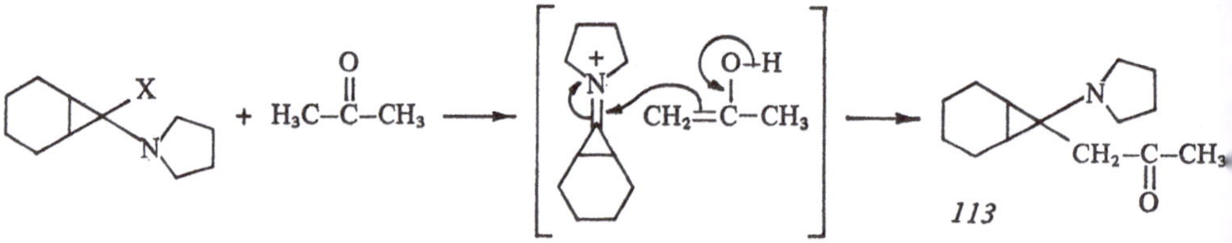

54, X = N

56, X = OH

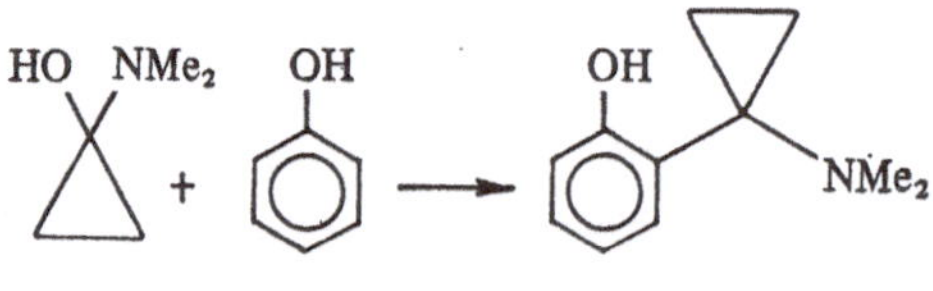

114

A particularly useful example of these alkylations is found in the ready reaction of nitroalkanes with 1,1-diaminocyclopropanes in the presence of methyl iodide. [93] The addition products formed as illustrated in Scheme 21 may be reduced to the amines and then deaminated to yield cyclobutanones by ring enlargement.

116

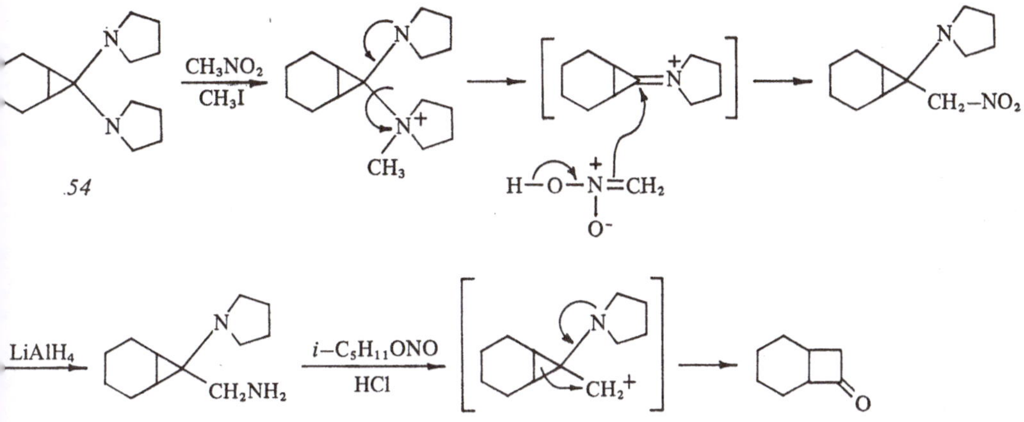

Scheme 21

4.1.5. Decarbonylation

Carbon monoxide may be eliminated from cyclopropanones either by thermal or photochemical processes (Table 12). In fact, decarbonylation is sometimes an undesirable side reaction in the synthesis of cyclopropanones. For example, the reaction of dimethylketene and ethyl diazoacetate affords carbon monoxide and ethyl acrylate *115* rather than the desired ketone. [20]

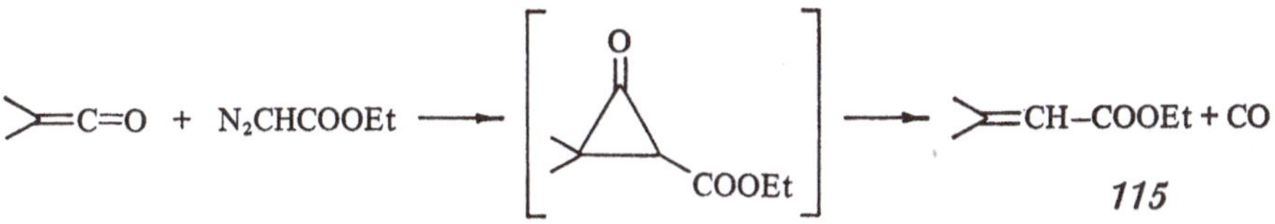

Similarly, photolysis of the cyclobutane-1,3-diones *19—21* produces varying amounts of ethylenes, ketenes and polymers in addition to carbon monoxide and the cyclopropanone, as was shown in Scheme 4, Section 2.3. [10,81]

Cyclopropanones may un dergo another type of thermal reaction which competes with decarbonylation, *i.e.*, cleavage of the C_2—C_3 bond. Thus, instead of losing carbon monoxide upon warming, the cyclopropanone, *31* appears to undergo ring opening to the oxyallyl species *116* which can be trapped as the furan cycloadduct *117*. [34] By contrast, photochemical decarbonylation of *31* is a facile reaction even at liquid nitrogen temperatures (Table 12). [34]

117

Table 12. Cyclopropanone decarbonylation reactions

Cyclopropanone	Conditions	Products	Ref.
	Thermal		

R = *t*—Bu

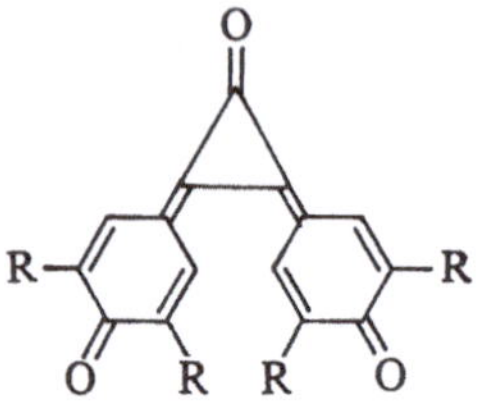

	Room temp.		131)
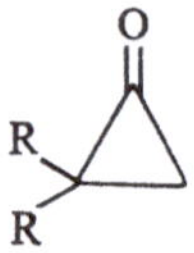	600 °C	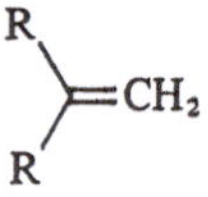	55a)
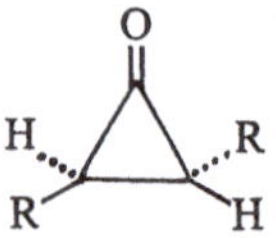	150 °C	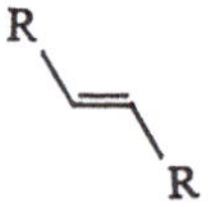	64)

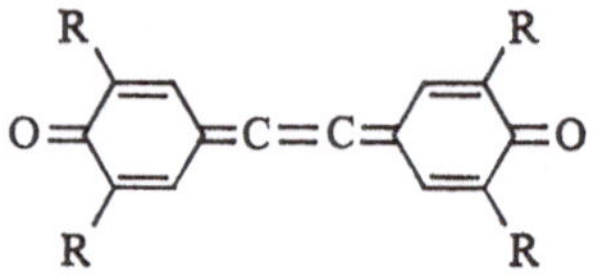

Photochemical

	Gas phase, 2920 Å 3650 Å	$H_2C=CH_2$ $H_2C=CH_2$  +	75)
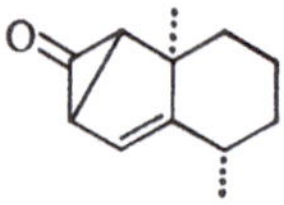	Neat glass, 3000—3600 Å − 190 °C	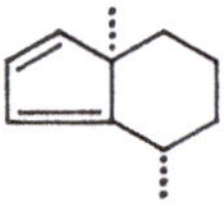	34)
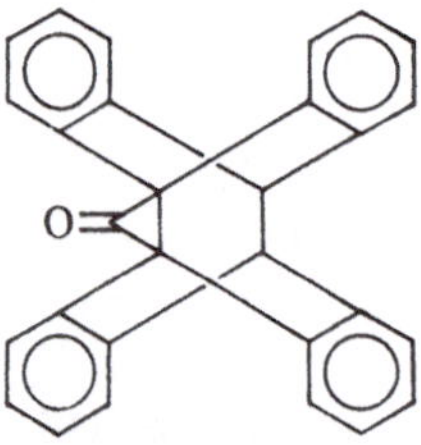	3500 Å, benzene	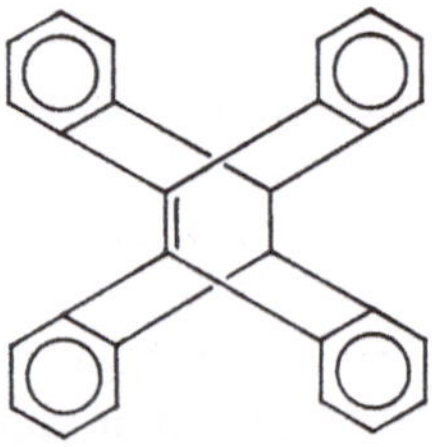	33)
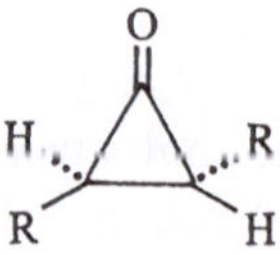	Sunlamp, 0 °C.	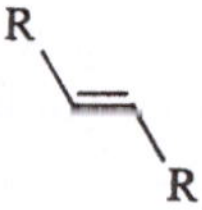	64)

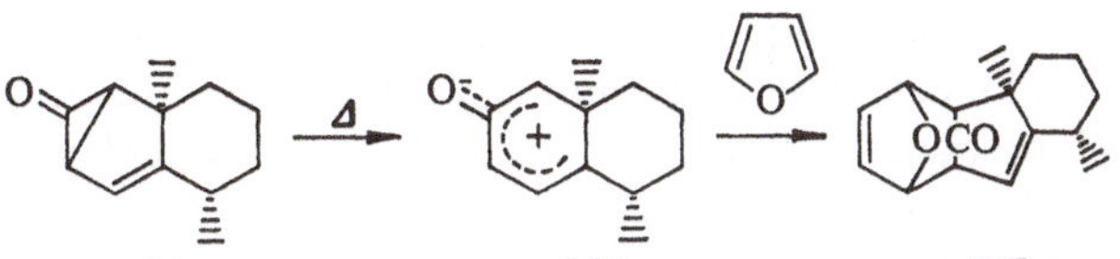

31   116   117

In a kinetic study of optically active *trans*-2.3-di-*t*-butylcyclopropanone (*52*), Greene and coworkers have been able to observe ring opening and decarbonylation as two distinct thermal reactions (Scheme 22). [64] Whereas the former process is measurable at 80 °C in terms of the rate of racemization, the latter process is not observed until 150 °C. That racemization does proceed through the oxyallyl species *118* is supported by such additional data as a first-order rate dependence and zero incorporation of solvent deuterium. The decarbonylation reaction is, of course, irreversible and the data are consistent with two pathways — a concerted elimination, or a stepwise fragmentation involving the intermediacy of a 1,3-diradical species.[g]

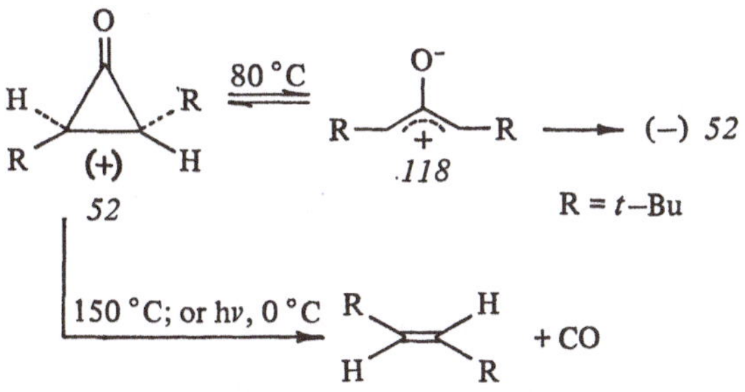

Scheme 22

Unlike the thermal reaction, irradiation of (+)*52* produces a loss of optical activity commensurate with decarbonylation. [64] Since the two rates are identical, photochemical decarbonylation apparently does not compete with or involve significant production of *118*, at least under these reaction conditions.

A recent photochemical study of cyclopropanone has demonstrated that two reactions take place at certain wavelengths. [75] Although the quantum yield for decomposition of 2 (ϕ_{cp}) is approximately unity over the range $\lambda = 2920$—3650 Å, the quantum yield for ethylene formation

---

[g] Although a diradical intermediate might be expected to give both *cis* and *trans* products, a high rotational barrier about the C_2—C_3 bond created by two *t*-butyl substituents may inhibit formation of the *cis* isomer.

($\phi_{C_2H_4}$) is wavelength dependent and varies from 1.0 at 2920 Å to 0.59 at 3650 Å.[h] The behavior of ϕ_{cp} and $\phi_{C_2H_4}$ together with the appearance of polycyclopropanone on the cell walls at longer wavelengths suggests that the n,π* excited singlet state of cyclopropanone (2*) is able to decompose in more than one manner as shown in Scheme 23. Thus, when 2* is in a lower vibrational state (~6 kcal/mol at 3650 Å), decarbonylation is sufficiently slow to allow an alternate mode of deactivation to compete. Since n,π* triplet states have been implicated in the photo-chemistry of cyclic ketones [94], it was proposed [75] that 2* undergoes intersystem crossing to a relaxed triplet state 2^t. Although α-cleavage to the biradical species is rapid in the n,π* excited triplet state of some ketones [95], 2^t may be sufficiently long-lived to form polymer on contact with the cell walls.

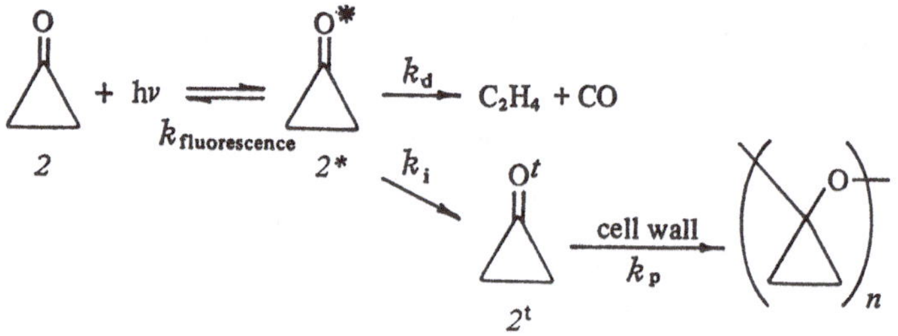

Scheme 23

4.2. Ring Expansion to Cyclobutanones and β-Lactams

Reactions of cyclopropanones with nucleophiles frequently lead to ring enlargement reactions since the formation of four-membered rings from the reactive intermediates is accompanied by a considerable reduction in strain energy. Thus, 2 reacts with diazomethane to form cyclobutanone[96], with hydrazoic acid to form β-lactam [76,89] and, under special condi-tions, with amines and hydroxyl amine derivatives to form N-sub-stituted β-lactams [87] (Scheme 24).

Cyclobutanones may be prepared in a one-step procedure, i.e., without isolating the intermediate cyclopropanone, simply by adding the ketene to excess diazoalkane. [97,98] That cyclopropanones are inter-mediates has been established by carbon-14 labeling studies [99] and

---

[h] The photolysis of ketene produces similar quantum yields supporting the prop-osition that the principle reactions involve the formation and decomposition of cyclopropanone (Kistiakowsky, G. B., Sauer, K.: J. Am. Chem. Soc. 78, 5699 (1956)).

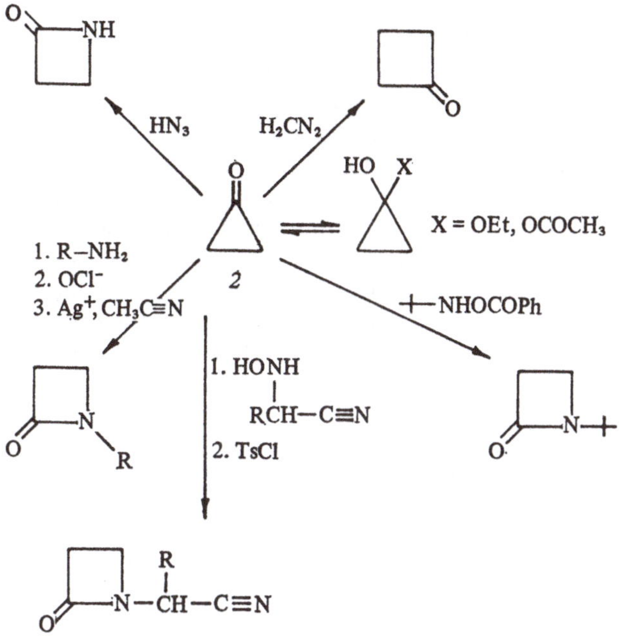

Scheme 24

by a comparison of product distributions in the diazoalkane-ketene and the corresponding diazoalkane-cyclopropanone reactions.[96] The reaction requires only 30 min. at $-78\,°C$ and, under these conditions, cyclobutanones are inert to further ring enlargement. When the cyclopropanone precursor is unsymmetrically substituted, the reaction with diazomethane leads to a mixture of cyclobutanones (Table 13a).[96] Other examples of cyclopropanone to cyclobutanone conversion *via* diazoalkanes are listed in Table 13b.

In a very different process, 2,2-di-*t*-butylcyclopropanone *66* may be converted to the cyclobutanone *119*.[55a] The reaction is acid-catalyzed and may proceed by the migration of a methyl group in the ring-opened species. Conversion of *66* to *119* is slow, even under vigorous conditions, *i.e.*, refluxing *66* with excess acetic acid in chloroform for three days.

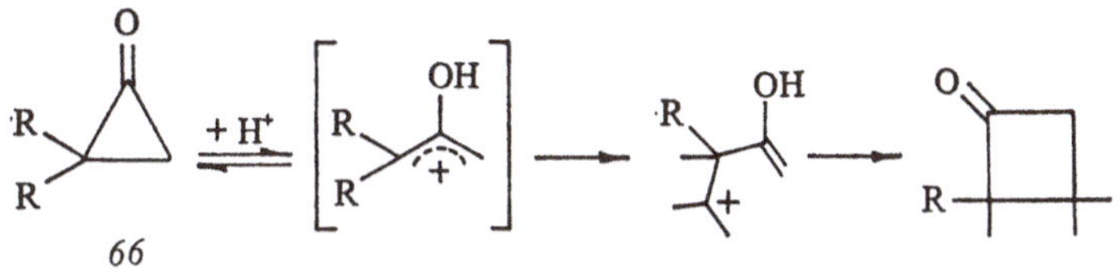

R = C(CH₃)₃

$R = C(CH_3)_3$

121

It appears that this route to cyclobutanones is applicable only to the most stable as well as the most favorably substituted cyclopropanones.

One of the most common ring enlargements in the chemistry of cyclo-propanones is observed in connection with alkylation reactions in which a double or triple bond is introduced in conjugation with the 3-membered ring. [76] Thus, as discussed in Section 4.1.4, 1-vinylcyclopropanol is a useful precursor of the cyclobutanone system, since, with a variety of electropositive reagents, it may be transformed to the hydroxycyclo-propyl carbinyl system and thence to the cyclobutanone (Scheme 20). A similar type of ring enlargement takes place in the case of 1-ethynyl-cyclopropanol *120*, forming the unsaturated cyclobutanone *121*.

Table 13a. Cyclobutanones from ketene-diazomethane reactions [96]

Ketene	Cyclopropanone intermediate	Cyclobutanones (product ratio, %)	
⊢CH=C=O		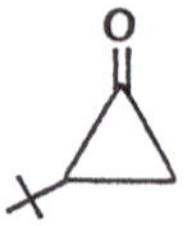 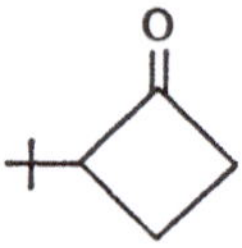 (35)	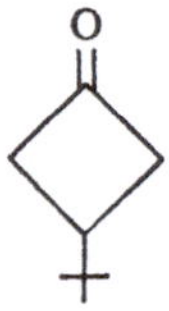
 (CH₃)₂C=C=O	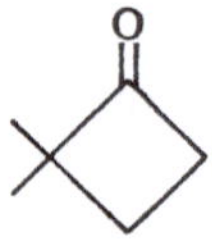	(36)	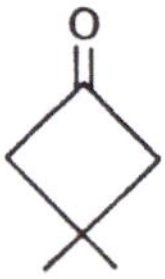
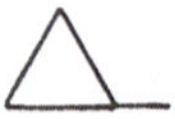 —CH=C=O		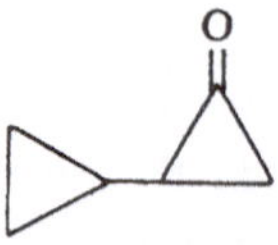 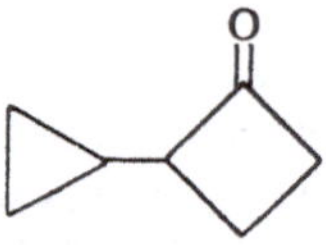 (35)	
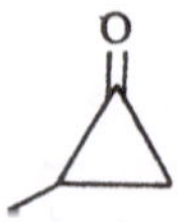 CH₃CH=C=O	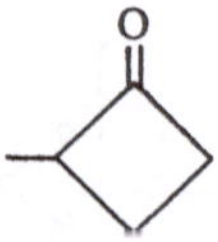	(43)	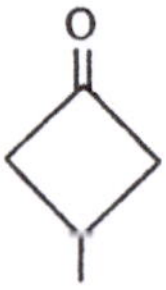

Table 13b. Cyclopropanone-cyclobutanone conversions *via* diazoalkanes [96)]

Cyclopropanone	Diazoalkane	Cyclobutanones (product ratio, %)			
	CH_3CHN_2	(12)	(31)	(40)	(17)
	CH_2N_2	(37)		(63)	
	CH_3CHN_2	(70)		(30)	

120 *121*

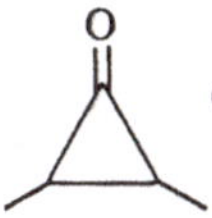

Ring enlargement of cyclopropanones appears to be a very promising method for generating the biologically important β-lactam system. [76,89)] Early studies showed that cyclopropanone precursors on treatment with azide ion at pH 5.5 underwent rearrangement to the β-lactam system with varying efficiency. A particularly high yield example was found in the case of the fused ring system *122* (Table 14).

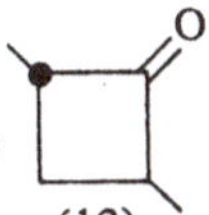

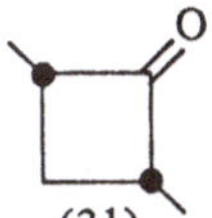

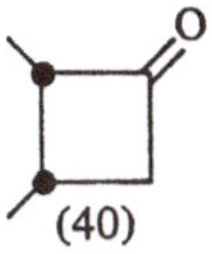

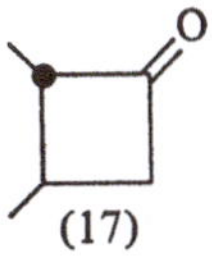

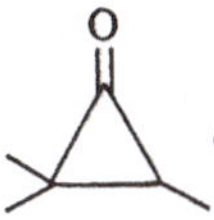

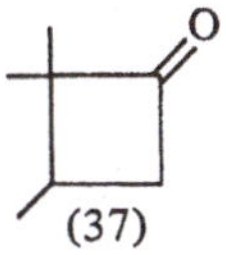

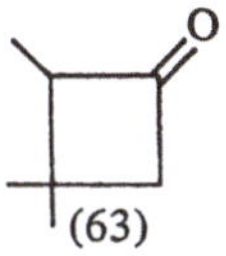

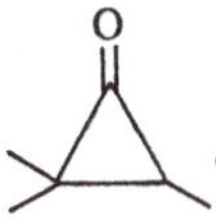

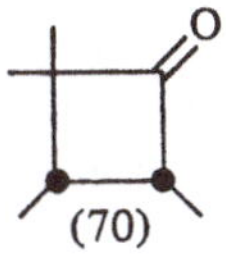

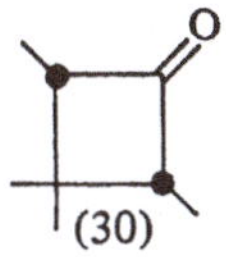

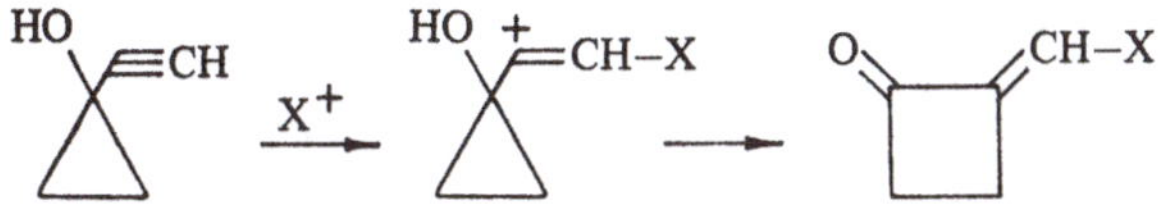

An extension of this reaction leading to a general synthesis of N-substituted β-lactams involves the addition of a primary amine to a freshly prepared solution of cyclopropanone, conversion of the resulting carbinol amine to the N-chloro derivative, and then decomposition of this intermediate with silver ion in acetonitrile. [87a)] The method permits one to prepare N-substituted β-lactams of great variety (Table 14), including those constructed from amino acid esters. [87b] The use of valine ethyl ester (*123*) as a nitrogen source leading to *124* is illustrated.

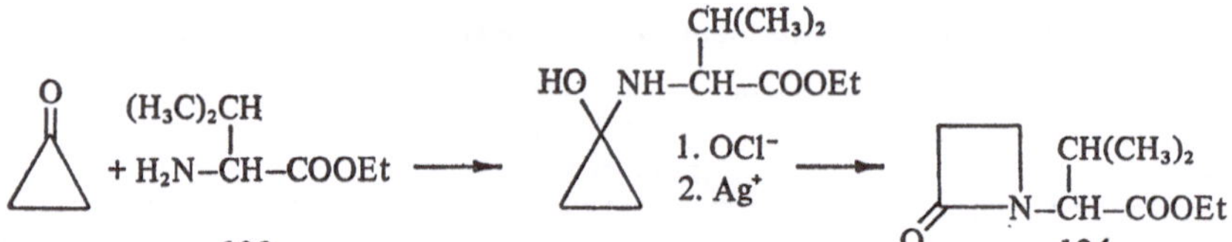

In both methods described above, the driving force for the ring enlargement is, most probably, the generation of a nitrenium ion adjacent to the cyclopropyl system (Scheme 25), which leads to rearrangement analogous to the cyclopropyl carbinyl-cyclobutyl conversion (Scheme 20). In the first case, the leaving group generating the electron deficient nitrogen at the 1-position is N_2; in the second case, it is halide. A further modification of the procedure permits introduction of the leaving group (OTs) under milder conditions. [100] Thus, addition of a substituted hydroxylamine to the cyclopropanone generates a N-hydroxy carbinol amine which may be converted to the nitrenium ion through the tosylate (Scheme 25). We found this procedure to be the most general and mildest of the three variations as summarized in Table 14.

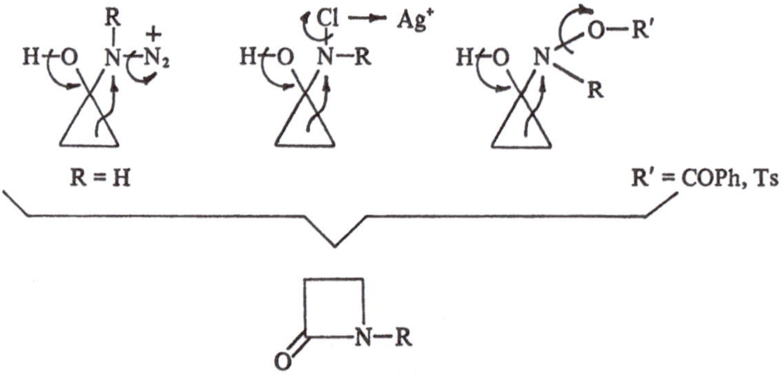

Scheme 25

4.3. Ring Opening Reactions

The special geometric requirements of the three-membered ring, resulting in bent and weakened bonds, renders the cyclopropane ring highly susceptible to both homolytic and heterolytic cleavage. Since each bond is labile, the site of cleavage depends upon the attacking species and, to a lesser degree, upon the ring substituents. In general, acid catalyzed ring opening usually involves the C_2-C_3 bond while other electrophiles, nucleophiles, and oxidizing agents cause rupture of the C_1-C_2 bond.

Table 14. β-lactams from cyclopropanones

Cyclopropanone source	Nitrogen source, conditions	β-lactam	Yield,%	Ref.
HO OEt	NaN₃ pH 5.5		(21)	76)
X = Y = N⟨⟩ X = OH; Y = N⟨⟩ X = OH; Y = OEt	NaN₃ pH 5.5	122	(63) (67) (95)	89)
	1. RNH₂ 2. OCl⁻ 3. Ag⁺			
		⟨⟩ (cyclohexyl)	(61)	
		n—Bu	(43)	87a)
		sec—Bu	(38)	
		t—Bu	(52)	
		—CH₂COOEt	(33)	
		CH₃CHCOOEt	(47)	
		(CH₃)₂CHCHCOOEt	(65)	87b)
		(CH₃)₂CHCH₂CHCOOEt	(65)	
		PhCH₂CHCOOEt	(70)	
	Ph—C(=O)—O—NH—t—Bu		(40)	87a)
H—O OCOMe	1. HO—NH—CH—C≡N (R) 2. TsCl			
		R CH₃CH₂CH₂—	(45)	
		(CH₃)₂CH—	(41)	
		CH₃CH₂CH₂CH₂—	(40)	100)
		(CH₃)₂CHCH₂—	(45)	
			(43)	
	HO—NHPh			

4.3.1. Cleavage with Base. The Favorskii Rearrangement

The cleavage reaction of cyclopropanones with base is synonymous with the second half of a normal[1] Favorskii rearrangement (Scheme 8, Section 2.5). The parallel course of these two reactions has been demonstrated by labeling [49], trapping [13], and product studies. [101] In accordance with these experimental findings, cyclopropanone reacts with base as shown in Scheme 26. The anionic carbonyl adduct *125* may also be generated from cyclopropanone hemiacetals by proton abstraction with base. Cleavage of *125* yields the carbanion *126* which affords the product *127* upon proton capture. Some illustrative reactions are given in Table 15.

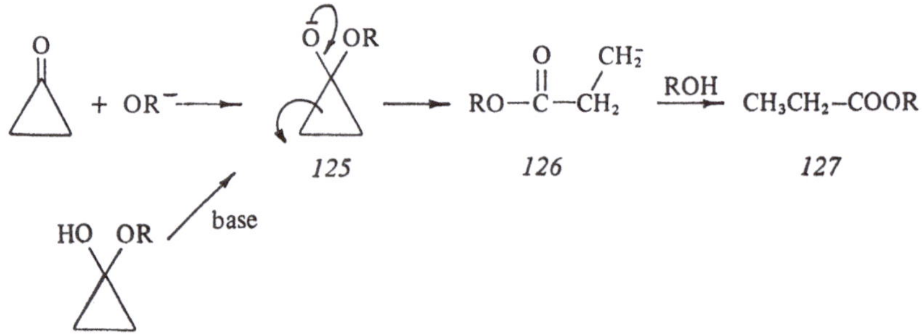

Scheme 26

When the cyclopropanone is unsymmetrically substituted, the C_1—C_2 and C_1—C_3 bonds are no longer identical and one would expect that the two resulting products would be formed in a ratio reflecting the relative stabilities of the C_2 and C_3 carbanions. Although this consideration provides a good working hypothesis, there are examples (Table 15) where the product distribution does not parallel the usual order of carbanion stability, *i.e.*, $1° > 2° > 3°$. In the reaction of 2,2-di-*t*-butyl-cyclopropanone, the predominant product is *not* derived from the primary carbanion, but rather the tertiary carbanion. [55a] In this case, the lability of the C_1—C_2 bond is apparently increased by the steric congestion resulting from the presence of two *t*-butyl substituents at C_2. [101]

Base-induced ring opening may also lead to stereospecific ester formation. Stereoselectivity with retention of configuration suggests that the carbanion formed upon cleavage does not undergo inversion prior

---

[1] When abstraction of the α-hydrogen by base is difficult, if not impossible, reaction occurs by a benzylic acid-like rearrangement (Warnhoff, E. W., Wond, C. M., Tai, W. T.: J. Am. Chem. Soc. *90*, 514 (1968) and references therein).

Table 15. Cleavage reactions of cyclopropanones with base

Cyclopropanone or precursor	Base	Products R=t–Bu	Isomer ratio	Total yield, %	Ref.
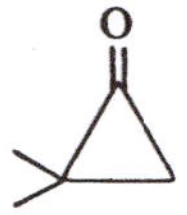	MeONa	(CH$_3$)$_3$CCOOMe		100	103)
	MeONa	(CH$_3$)$_3$CCOOMe			103)
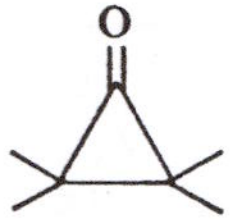	MeONa	(CH$_3$)$_2$CHCHOC(CH$_3$)$_2$OMe + (CH$_3$)$_2$CHC(CH$_3$)$_2$COOMe	1:33		132)
	MeONa	(CH$_3$)$_2$CHC(CH$_3$)$_2$COOMe			132)
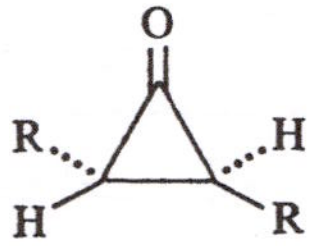	MeONa	threo—RCH$_2$CH(R)COOMe		97	102)

Table 15 (continued)

Cyclopropanone or precursor	Base	Products R=t–Bu	Isomer ratio	Total yield, %	Ref.
[cyclopropanone structure]	MeONa	$CH_3CH_2C(CH_3)_2COOMe + (CH_3)_2CHCH(CH_3)COOMe$	4:1	100	101)
[structure with HO OMe]	MeONa	$CH_3CH_2C(CH_3)_2COOMe + (CH_3)_2CHCH(CH_3)COOMe$	4.5:1	100	101)
[cyclopropanone structure]	t-BuOK	$CH_3CH_2C(CH_3)_2COOR$		90	101)
[structure with HO OR]	t-BuOK	$CH_3CH_2C(CH_3)_2COOR$		90	101)
[structure with R R]	MeONa	$R_2C(CH_3)COOMe + R_2CHCH_2COOMe$	1:3		55a)

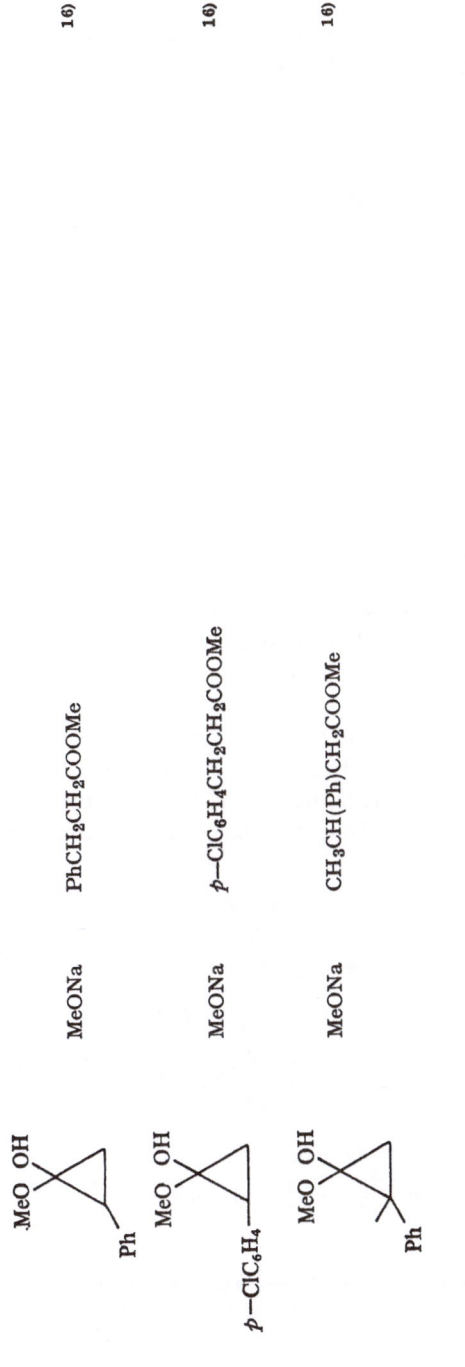

to proton capture. As an example, *trans*-2,3-di-*t*-butylcyclopropanone reacts with sodium methoxide or ethylene glycolate in the corresponding deuterated alcohol affording the *threo* ester *128*. [102]

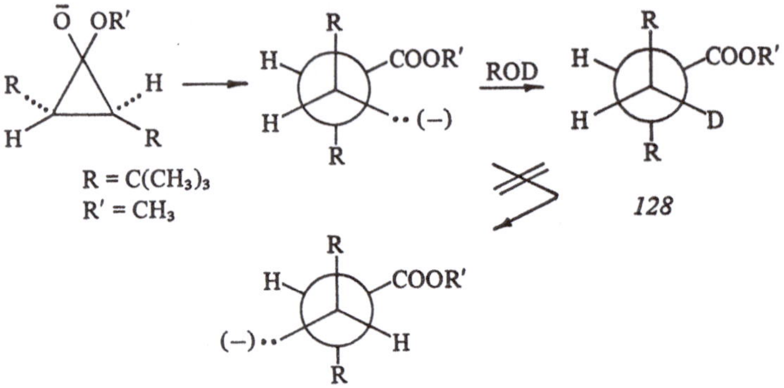

$R = C(CH_3)_3$
$R' = CH_3$

128

4.3.2. Ring Opening Under Neutral and Acidic Conditions

Cyclopropanones and their labile derivatives (*e.g.*, hemiacetals) tend to undergo ring opening of the C_2–C_3 and/or C_1–C_2 bond(s) forming propanones or propionates. [1,5] The decomposition process is slow in neutral media but is accelerated under mildly acidic conditions. As shown in Scheme 27, the choice of cleavage site is dependent upon the relative stabilities[J] of the cyclopropanone-HX adduct *129* compared to the hydroxyallyl cation *130*. In general, cleavage of the C_2–C_3 bond is favored when the ring carbons are bound to substituents capable of stabilizing the hydroxyallyl cation, *i.e.*, electron-donating groups.

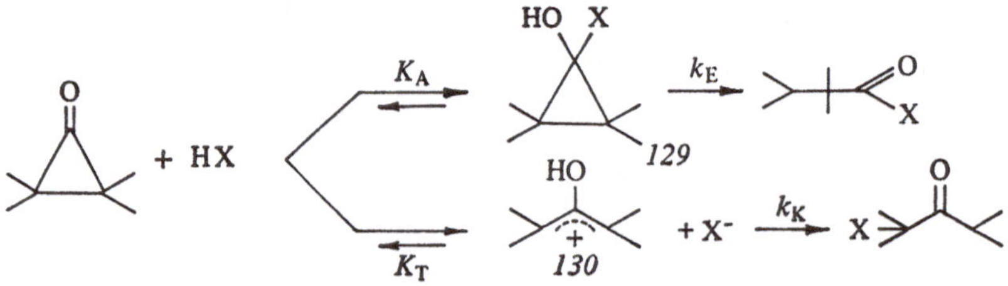

Scheme 27

(A = adduct; T = tautomer; E = ester; K = ketone.)

[J] As used here, the term relative stabilities includes both kinetic and thermo dynamic considerations, that is, the rate constants k_K and k_E, and the equilibrium constants K_T and K_A.

In accord with the above, substituent effects in 2-arylcyclopropanone herniketals indicate that initial conversion to the parent ketone takes place prior to a C_2—C_3 ring opening. [16] Although hemiketal-ketone equilibration is slow under neutral conditions, ketone formation and hence cleavage through *130* is accelerated in acid.

In Table 16, cleavage products under different conditions are given for several cyclopropanones. 1-Methoxy-2-phenylcyclopropanol [16], 2,2-dimethylcyclopropanone [5,103] and 2,2-di-*t*-butylcyclopropanone [55a] yield ketonic products exclusively suggesting the presence of favorable substituent interactions in the allyl cation. On the other hand, under neutral conditions, C_1—C_2 and C_2—C_3 cleavages are competitive in tetramethylcyclopropanone hemiacetals. [8,104] In the absence of acid, conversion to the ketonic form is slow and 1,2-ring opening becomes significant. In the case of the parent cyclopropanone and its hemiacetals [1,5], formation of the unsubstituted hydroxyallyl cation (C_2—C_3 cleavage) is not favored.

Cyclopropanone acetals require far more vigorous conditions (concentrated acid and heating) for ring cleavage compared to hemiacetals. As shown in Scheme 28, the reaction may proceed in two directions, one involving O-protonation (*a*) and the other C-protonation (*b*). In the case of 1,1-diethoxycyclopropane where both paths are competitive, refluxing hydrochloric acid yields both chloroacetone and ethyl propionate (Table 17). [25]

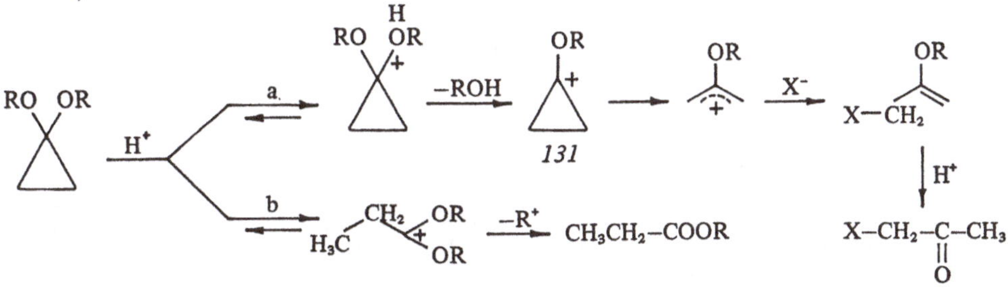

Scheme 28

The spiro ketal derivatives studied by Giusti [105] are remarkably uniform in their behavior toward acid hydrolysis (Table 17). Ring opening usually occurs at the site of the less-substituted carbon and, in all cases, the ester is the sole product. The overwhelming preference shown by cyclopropanone ethylene ketals for path *b*) in Scheme 28 may be attributed to (*a*) the reversibility of path *a*) due to rapid intramolecular ketalization at the incipient C_1-carbonium ion *131* (R=CH_2CH_2OH) and (*b*) the stability of the intermediate dioxocarbonium ion *131 a* generated in path *b*). [106]

131

H. H. Wasserman, G. M. Clark, and P. C. Turley

Table 13. Decomposition of cyclopropanones and cyclopropanone hemiacetals under acidic and neutral conditions

Cyclopropanone or precursor	Conditions	Products	Ref.
HO OR (cyclopropane)	Standing, or acid	CH_3CH_2COOR R = H, CH_3, C_2H_5	1,15)
	HX	X + O=C—X X = Cl, OAc	5,103)
O=(cyclopropane with isopropyl)	vpc	(enone) + (methyl vinyl ketone)	103)
	Standing, or refluxing R'OH	COOR' + R'O—C(=O)— R = CH_3, C_2H_5 i-C_3H_7 R' = CH_3, C_2H_5	9a,104)
HO OR (tetramethyl cyclopropane)	glc	COOR + RO—C(=O)—C R = C_2H_5	9a)
	85°/N₂	Above products + COOR + ROH R = CH_3	104)

	Heating/O_2	Above products + 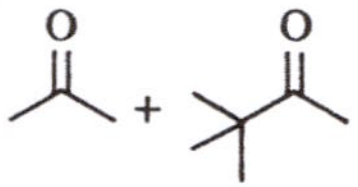	R = CH_3	104)
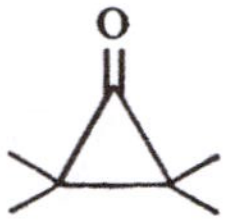	vpc	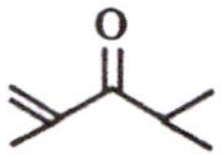		8)
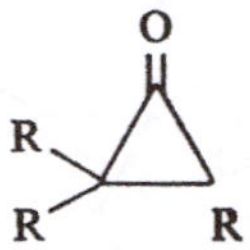 R = $C(CH_3)_3$	CH_3COOH	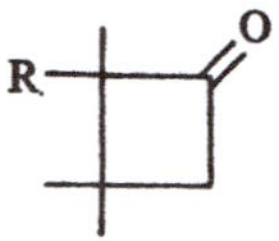		55a)
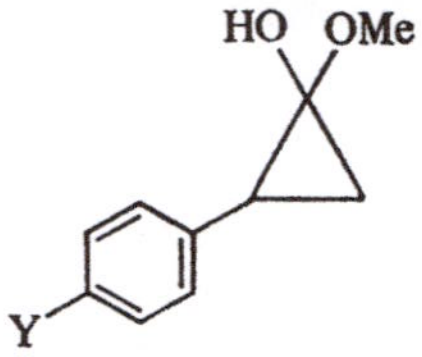	HX	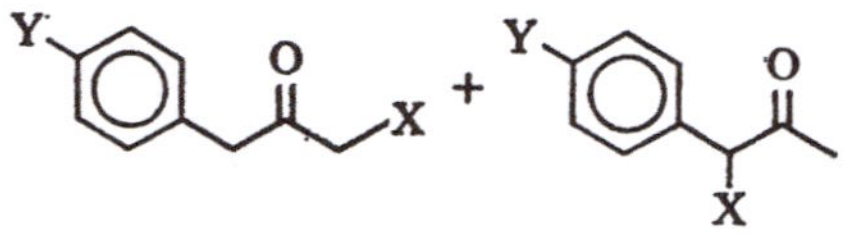	Y = H, Cl	16)

Table 17. Ring opening reactions of cyclopropanone acetals under acidic conditions

Acetal	Conditions	Products	Yield, %	Ref.
EtO OEt (cyclopropane)	Conc. HCl, reflux	CH₃CH₂COOEt	12	25)
		CH₃CH₂COOH	30	25)
MeO OMe, Cl Cl, CH₃ (cyclopropane)	Conc. HCl, reflux	ClCH₂—C(=O)—CH₃	34	
		(CH₃)₂C=C(Cl)—COOR R=H, CH₃	66	25)
 MeO OMe (dimethylcyclopropane)	Conc. HCl, 50°, 1 h	Cl—C(CH₃)₂—C(=O)—CH₃	85	18)
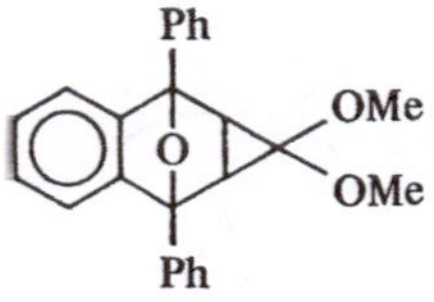 Ph, OMe OMe, Ph (anthracene acetal)	CF₃COOH, warming	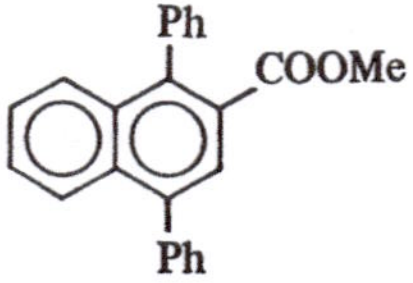 naphthalene (Ph, COOMe, Ph)		61)
 SMe SMe (bicyclic cyclohexane)	HCOOH, reflux, 20—30 h	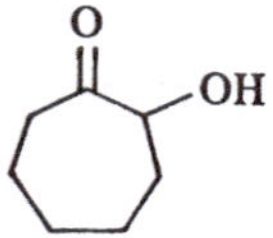 cycloheptanone-OH	56	110)

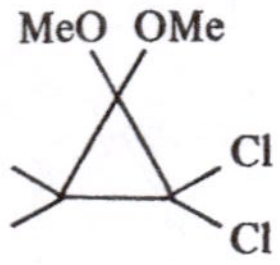

134

Reagent/Conditions	Product	Yield (%)	Ref.
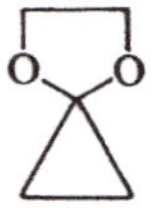 HCl/CCl$_4$[1], room temp.	C$_2$H$_5$COCH$_2$CH$_2$Cl	90	105)
C$_2$H$_5$COOH[1], warming	C$_2$H$_5$COCH$_2$CH$_2$OCC$_2$H$_5$	83	105)
 HCl/CCl$_4$[1], room temp.	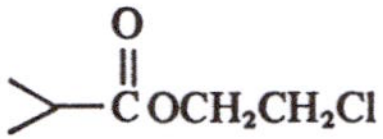 COCH$_2$CH$_2$Cl	93	105)
 Cl$_2$CHCOOH, ether, 0 °C	 COCH$_2$CH$_2$OCCHCl$_2$	90	105)
HCOOH/ether[1], 0 °C	 COCH$_2$CH$_2$OCH	65	105)
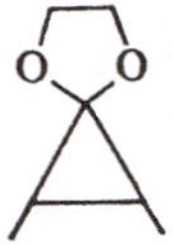 HBr/CCl$_4$[1]	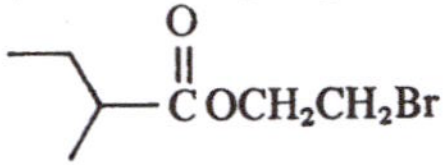 COCH$_2$CH$_2$Br	90	105)
HCl/CCl$_4$[1]	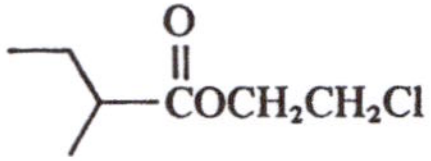 COCH$_2$CH$_2$Cl	90	105)
C$_2$H$_5$CO$_2$H[1], warming	 COCH$_2$CH$_2$OCOC$_2$H$_5$	85	105)
HCl/CCl$_4$[1], room temp.	PhCH$_2$CH$_2$COCH$_2$CH$_2$Cl	34	105)
	CH$_3$–CH–COCH$_2$CH$_2$Cl, Ph	66	

1) Reagents are anhydrous.

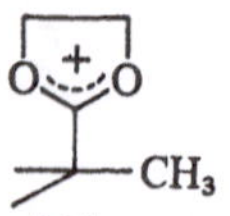

$$131a$$

The acid hydrolysis of cyclopropanone ethylene ketals to the corresponding esters appears to be a facile, high yield reaction (75—90%) which may be of synthetic importance. Thus, as shown in Scheme 29, a ring contraction similar to a Favorskii rearrangement may be achieved under mildly acidic conditions.[107]

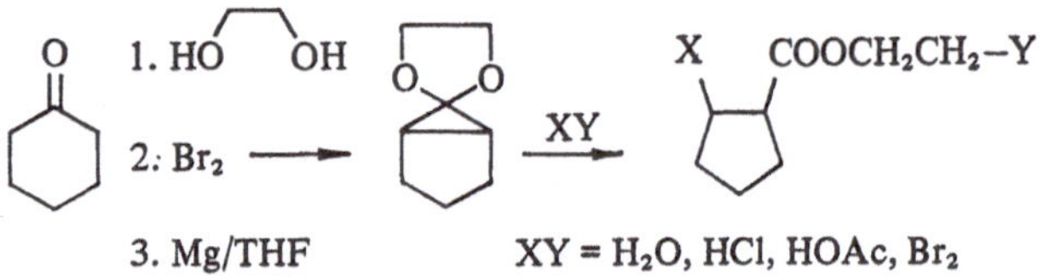

$$
\begin{array}{cc}
\text{1. HO} \quad \text{OH} & \text{X} \quad \text{COOCH}_2\text{CH}_2-\text{Y} \\
\text{2: Br}_2 & \xrightarrow{\text{XY}} \\
\text{3. Mg/THF} & \text{XY} = \text{H}_2\text{O, HCl, HOAc, Br}_2
\end{array}
$$

Scheme 29

The nitrogen analogs of cyclopropanones also undergo acid-catalyzed ring opening (Scheme 30). When the cyclopropanimine 57 is treated with one mole of deuterosulfuric acid in methylene chloride, C_2—C_3 cleavage occurs affording the zwitterion 132 (90%).[108] However, when the acid is a hydrogen halide, the two α-haloimines 133 and 134 are obtained with 133 the predominating isomer in all cases.

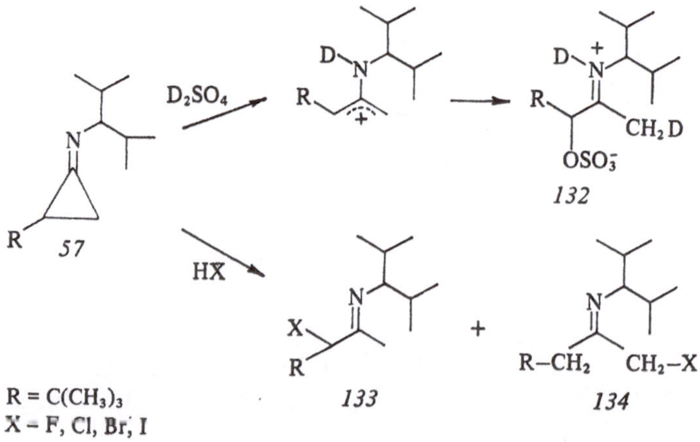

$$
\begin{array}{c}
\text{OSO}_3^- \\
132
\end{array}
$$

$$
R = C(CH_3)_3
$$
$$
X = F, Cl, Br, I
$$

$$133 \qquad 134$$

Scheme 30

4.3.3. Other Electrophilic Cleavage Reactions

The cyclopropanone ring is susceptible to attack by electrophilic reagents other than acids, *e.g.* bromine, phenol, acid chlorides. Most of these reactions have been observed in cyclopropanone acetals and all proceed by attack at C_2 or C_3 in a manner analogous to path *b*, Scheme 28. As shown in Table 18, esters are usually obtained but, under special bromination conditions, 1,1-dialkoxycarbonium halides are formed.[109] These salts lose ROBr upon warming to $-30\,°C$ to give the expected esters.

A cyclopropanone thioacetal has also been observed to undergo $C_1—C_2$ cleavage under the normal conditions for thioacetal solvolyses.[110] Thus, the esters *135a* and *135b* are formed when the 1,3-dithiopropane ketal of 7,7-norcarane is reacted with mercuric chloride. In this case, $HgCl^+$ acts as an electrophile and attacks the three-membered ring. However, under similar conditions, the cyclopropanone methyl thioketal *136* forms the mixed ketal *137*. While the authors consider this result to represent an unusual example of a nucleophilic displacement at a cyclopropyl carbon atom [110], the reaction mechanism may involve the inter-

135a   *135b*

136   *138*   *137*

mediate ion *138* and the mechanism may be similar to that proposed for the silver nitrate-assisted methanolysis of the dibromide *139*.[111]

139

Table 18. Electrophilic reactions of cyclopropanone acetals

Acetal	Electrophile	Products, yields % and reaction conditions	Ref.
 EtO OEt △	Br$_2$/CCl$_4$	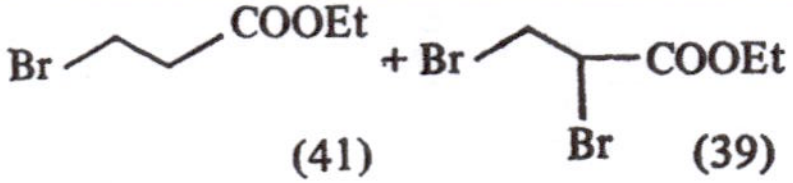 Br⌒⌒COOEt (41) + Br⌒⌒COOEt Br (39)	25)
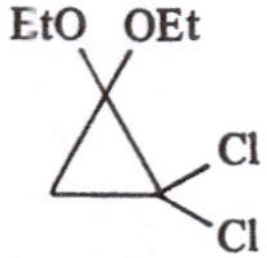 EtO OEt △ Cl Cl	Br$_2$/CCl$_4$	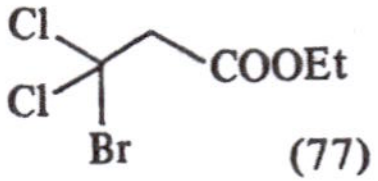 Cl Cl COOEt Br (77)	25)
 MeO OMe △	Br$_2$/CCl$_4$	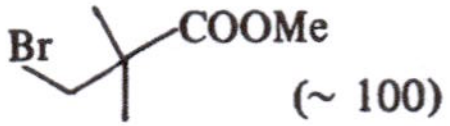 Br⌒COOMe (~100)	18)
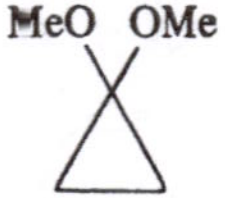 MeO OMe △	Br$_2$/SO$_2$	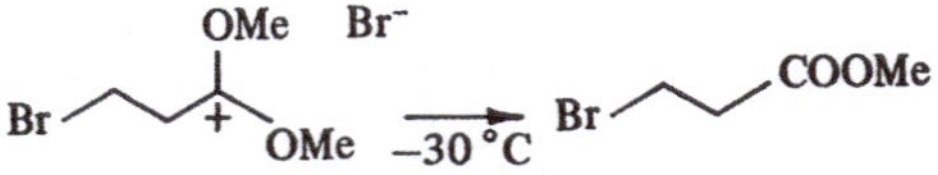 OMe Br$^-$ +OMe Br $\xrightarrow{-30\,°C}$ Br⌒⌒COOMe	109)
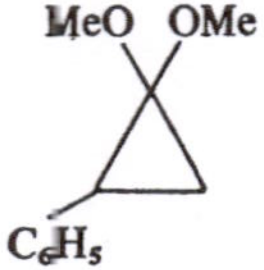 MeO OMe △ C$_6$H$_5$	Br$_2$/SO$_2$	 C$_6$H$_5$ OMe Br$^-$ +OMe Br $\xrightarrow{-30\,°C}$ C$_6$H$_5$ Br COOMe	109)

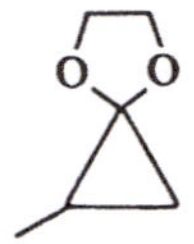

105)

105)

105)

105)

ClOH/H₂O, 5°

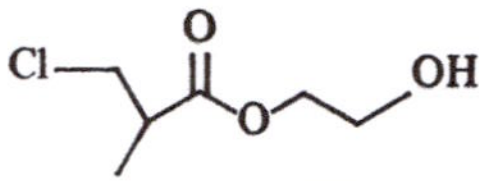

OH

(88)

n-C₄H₉SH

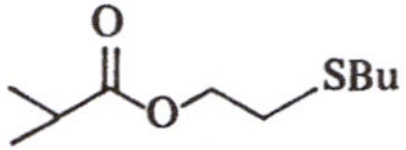

SBu   180 °C, 24 h

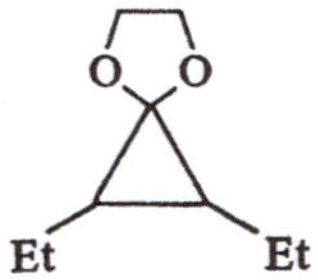

(90)

C₆H₅OH

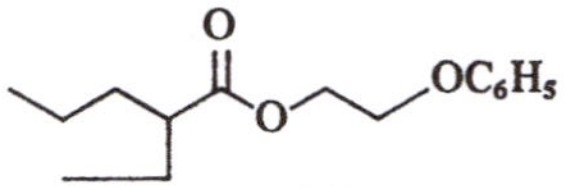

OC₆H₅   100 °C, 24 h

(50)

H₂O, H⁺

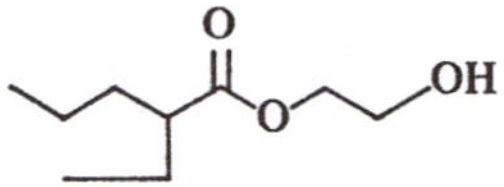

OH   100 °C, 2 h

(80)

Br₂/CCl₄

Br   0 °C

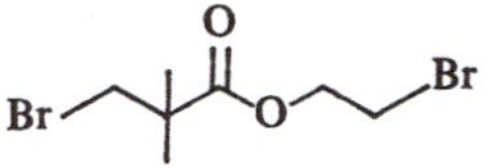

(85)

139

4.3.4. Oxidative Cleavage Reactions

A cyclopropane ring bound to an oxygen substituent as in cyclopropanol or cyclopropanone hemiacetal appears to be unusually susceptible to oxidative cleavage. The oxidizing agent may be a metal ion, *e.g.* Cu^{++} or Fe^{3+} [112–114], a peroxide [115a], or, in the case of alkyl-substituted systems, atmospheric oxygen.[115b,116] The reactions are often quite complex as would be expected of radical processes, but in all cases, only products derived from C_1–C_2 ring opening are obtained.

As Table 19 shows, the product composition resulting from metal ion oxidation is highly variable. The reaction course depends not only on the metal ion and its ligands, but also on the concentration and addition rate of the reactants. The key intermediates in these reactions are the β-propionate radicals, *e.g. 140*, which may be formed from highly unstable cyclopropyloxy radicals. The radical *140* has, in fact, been observed and characterized by ESR upon reaction of 1-methoxycyclopropanol with ceric sulfate.[112]

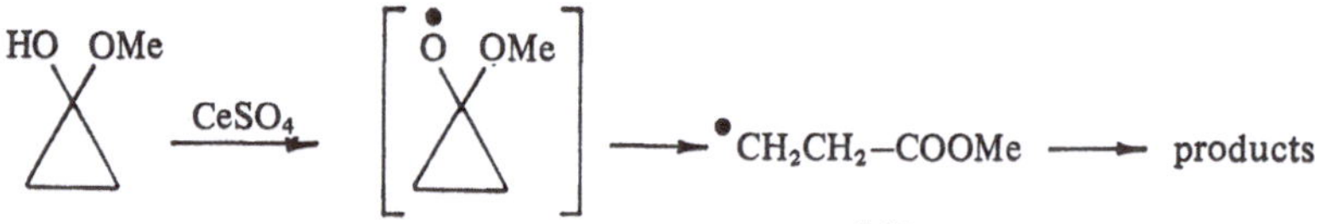

As indicated in Scheme 31, *140* may undergo several different reactions including addition to activated olefins in solution.[114] Radical oligomers such as *141* and *142* are ultimately reduced by the reduced form of the metal.

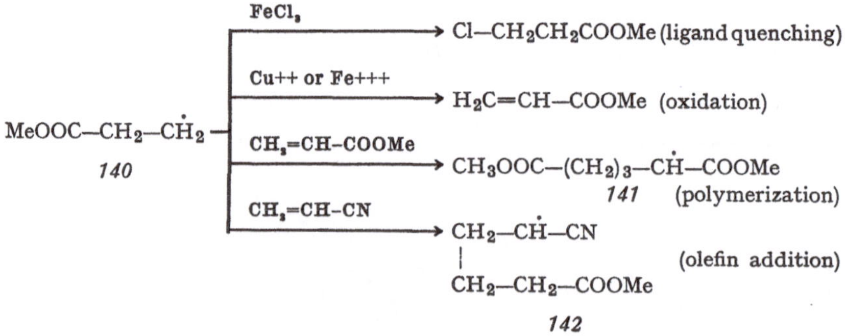

Scheme 31

When the cyclopropanone ring is unsymmetrically substituted, two radicals are produced if C_1—C_2 and C_1—C_3 cleavages are competitive. In the reaction of 1-methoxy-2,2-dimethylcyclopropanol with cupric ion, product studies indicate that approximately twice as much primary radical (143) is produced as tertiary (144) (Table 19). Although these two radicals would be expected to condense with methyl β-methylcrotonate (145), addition of 145 to the reaction mixture does not affect the nature of the products and the principal materials obtained, 146 and 147, (Table 19) probably arise by radical coupling.[113]

Alkyl substituted cyclopropanols and cyclopropanone hemiacetals [115,116a] also undergo oxidative cleavage when exposed to air or peroxides; the initial products are hydroperoxides such as 148. In the case of 1-methoxy-2,2-dimethylcyclopropanol, the reaction can be followed by observing the emission peaks in the NMR spectrum, and these CIDNP effects have enabled identification of radical intermediates.[115a] With di-t-butylperoxylate (TBPO), the isomeric radicals 143 and 144 are formed and these may undergo a diverse number of further reactions as indicated by the complex product mixture given in Table 20.

4.3.5. Other Homolytic Cleavages

Cyclopropyl nitrites are considerably more reactive than other alkyl nitrites and readily rearrange below ice-bath temperatures.[117] Thus, the nitrite ester of tetramethylcyclopropanone methyl hemiacetal is easily converted to the β-propionate radical 149 at − 80 °C and, in the presence

Table 29. Oxidative cleavage by metal ions

Reactants	Products	Ref.
 HO OMe ▷◁ + Ce(SO$_4$)$_2$	$\left[\cdot \text{CH}_2\text{CH}_2\text{COOMe} \right]$ *140*	112a)
 HO OMe ▷◁ + CuSO$_4$/NH$_3$	MeOOC(CH$_2$)$_4$COOMe + CH$_2$=CH–COOMe *156* *157*	112a) 113)

 HO OMe ▷◁ + Fe(NO$_3$)$_3$/H$^+$	*156* (51%) + *157* (10%) + MeOOC(CH$_2$)$_3$CH(CH$_2$)$_2$COOMe (14%)
	         │
	        COOMe
		113)

 HO OMe ▷◁ + FeCl$_3$/H$^+$	ClCH$_2$CH$_2$COOMe (97%)	112a) 113)

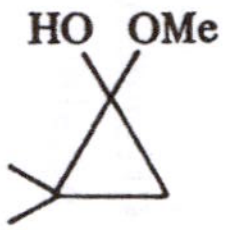

HO OMe ⟨structure⟩ + CuSO$_4$/NH$_3$ → (CH$_2$C(CH$_3$)$_2$COOMe)$_2$ (31%) + MeOOCCH$_2$C(CH$_3$)$_2$CH$_2$CH$_2$C(CH$_3$)$_2$COOMe (29%)
146
147
+ CH$_2$=C—CH$_2$COOMe (15%) + (CH$_3$)$_2$C=CH—COOMe (6%)
     |
     CH$_3$ 158   145
[113)]

HO OMe ⟨structure⟩ + Fe(NO$_3$)$_3$/H$_2$O → 158 (20%) + 145 (7%) + HO—C(CH$_3$)$_2$CH$_2$COOMe (43%)
                                                                        159
[113)]

HO OMe ⟨structure⟩ + FeCl$_3$/H$^+$ → ClC(CH$_3$)$_2$CH$_2$COOMe (73%) + 158 (3%) + 145 (1%) + 159 (1%)
[113)]

Table 20. Di-*t*-butylperoxalate oxidation of 2,2-dimethylcyclopropanone methyl hemiacetal [115a)]

Products and yields, %	
DCCl$_3$ as solvent, $t = 56\,°C$	C$_6$D$_6$ as solvent, $t = 65\,°C$

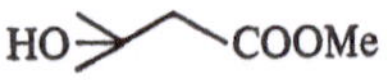

HO⟩⟨COOMe (5)   D⟩⟨COOMe (35)

D⟩⟨COOMe (37)   D⟩⟨COOMe (24)

D⟩⟨—COOMe (13)   HO⟩⟨COOMe (5)

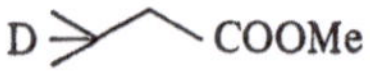

Cl⟩⟨COOMe (15)   ⟩=⟨COOMe (3)

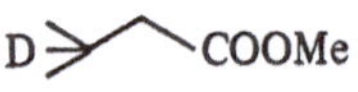

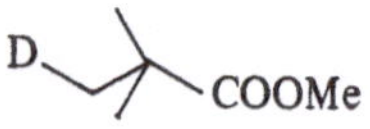

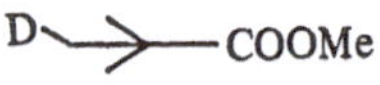

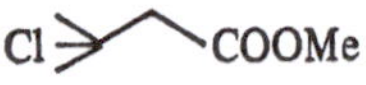

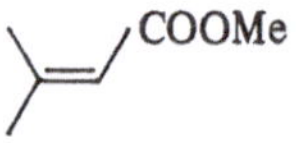

Cl$_3$C⟩⟨COOMe (5)   ⟩⟨COOMe (11)

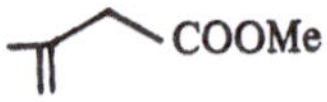

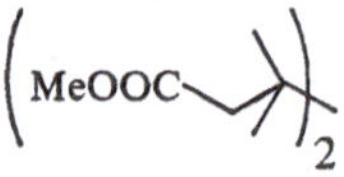

MeOOC⟩⟨⟨COOMe (7)

(MeOOC⟩⟨)$_2$ (7)

HO OMe   O=N—O OMe   [O=N⋯O OMe]

△ →NOCl→ △ → [△]

O=N—||—COOMe + Br—||—COOMe ←BrCCl$_3$— •—|—COOMe + NȮ

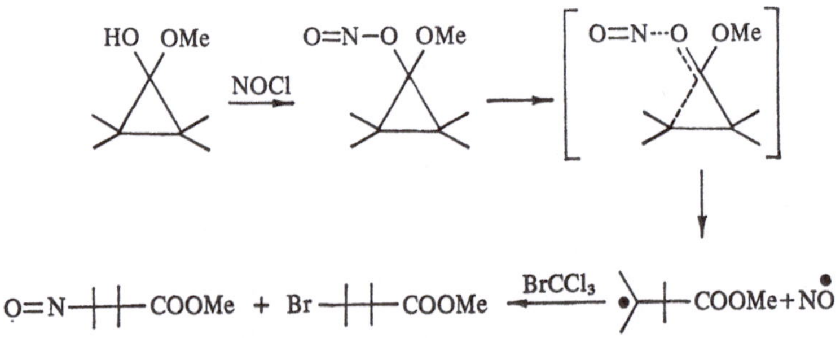

149

of bromotrichloromethane, affords methyl 3-bromo-2,2,3-trimethyl-butyrate (30%) and methyl 3-nitroso-2,2,3-trimethyl butyrate (40%). The remarkable facility of this reaction suggests a transition state in which there is considerable breaking of the C$_1$—C$_2$ bond as well as the N—O bond.

Finally, an example of a homolytic C_2—C_3 cleavage is provided by the catalytic hydrogenation of *trans*-2,3-di-*t*-butylcyclopropanone *(52)* to di-*neo*-pentyl ketone.[13]

4.4. Cycloaddition Reactions

In accord with their unusually high reactivity, cyclopropanones may undergo various types of cycloaddition reactions. In addition to Diels-Alder type addition with conjugated dienes ($4+3\rightarrow7$ cycloadditions), substituted cyclopropanones also react with aldehydes forming dioxolanes ($3+2\rightarrow5$), and with ketenes yielding β-lactones ($2+2\rightarrow4$).[118]

4.4.1. $4+3\rightarrow7$ Cycloadditions

The first example of a product from a $4+3\rightarrow7$ cycloaddition of a cyclopropanone was provided by Fort [119], who isolated the bicyclic ketone *37* from the reaction of α-chlorodibenzyl ketone with 2,6-lutidine in the presence of furan. Cookson and Nye [40] isolated *37* from the reduction of *bis*-α-bromobenzyl ketone with sodium iodide in the presence of furan, and also obtained the corresponding adduct with cyclopentadiene. The

latter reaction was pictured in terms of a zwitterionic intermediate, since no absorption in the 1800 cm^{-1} region of the infrared spectrum of the reaction mixtures could be detected.

Edelson and Turro [79] have examined the kinetics of the cycloaddition of 2,2-dimethylcyclopropanone with furan and with cyclopentadiene at 0 °C, as well as the relative reactivities of a number of alkyl-

145

cyclopropanones with these reagents (unsubstituted cyclopropanone does not react). The cycloaddition with furan and 2,2-dimethylcyclopropanone was found to be second order, first order in each reagent. Competitive reactions of various dienes with 2,2-dimethylcyclopropanone yielded the relative rates shown in Table 21.

Table 21. Relative rates of reaction of dienes with 2,2-dimethylcyclopropanone [79]

Diene	k Relative
	1
	~3
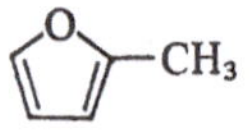 —CH$_3$	~1.25
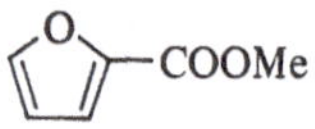 —COOMe	~0.37

Perhaps more informative are the data in Table 22 for the relative rates of reaction of various cyclopropanones with furan. The results are consistent with the mechanism given in Scheme 32 where $k_1 \gg k_2$ [furan], although the kinetic data do not distinguish between the closed cyclopropanone form and the zwitterionic intermediate. Increasing substitution would, of course, increase the stability of the allyl cation.

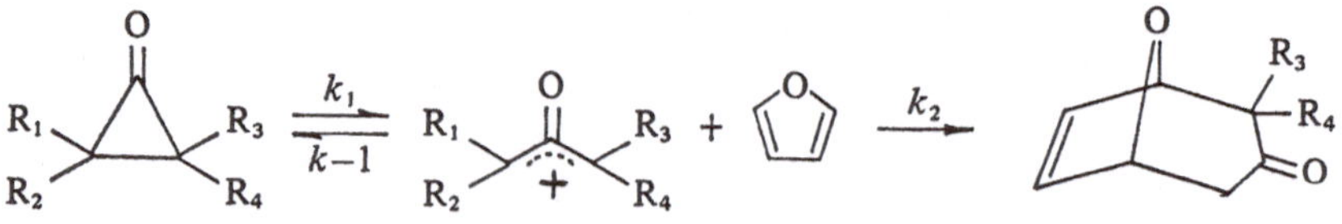

However, 2,2-dimethylcyclopropanone as well as the tetramethyl derivative have reactivities lower than expected from electronic considerations, and the authors suggest that steric effects associated with inward rotation of methyl groups in the transition state must also be considered. More recently, Hoffmann [120] has also suggested certain conformational

146

Table 22. Relative rates of reaction of
various cyclopropanones with furan [79]

Cyclopropanone	k Relative
	<0.0001
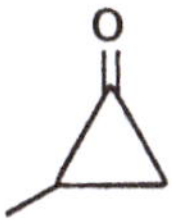	~1
	1
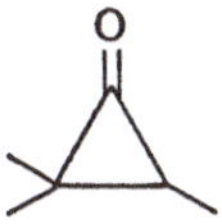	>100
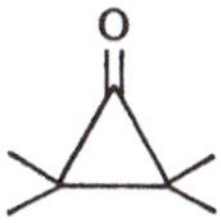	~1–2

preferences in the formation of intermediate oxyallylic cations in such
4 + 3→7 cycloaddition reactions.

Other 4 + 3→7 cycloaddition reactions of cyclopropanones with
dienes are listed in Table 23. Of particular interest is the reaction of
2,2-dimethylcyclopropanone with N-methylpyrrole to give the tropane
alkaloid *150*. [121]

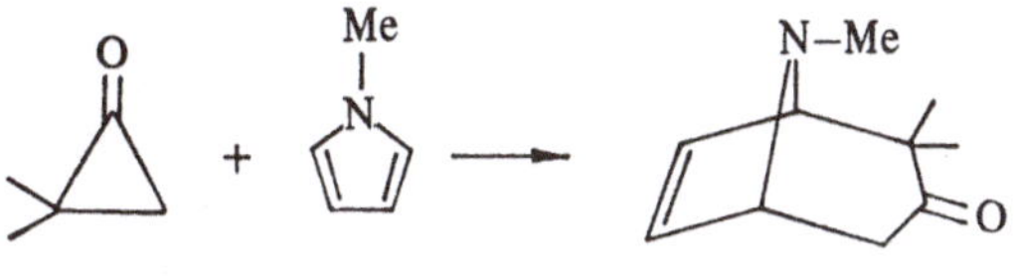

150

147

Table 23. 4 +3→7 Cyclopropanone cycloadditions [118,122)]

Cyclopropanone	Diene	Addition products

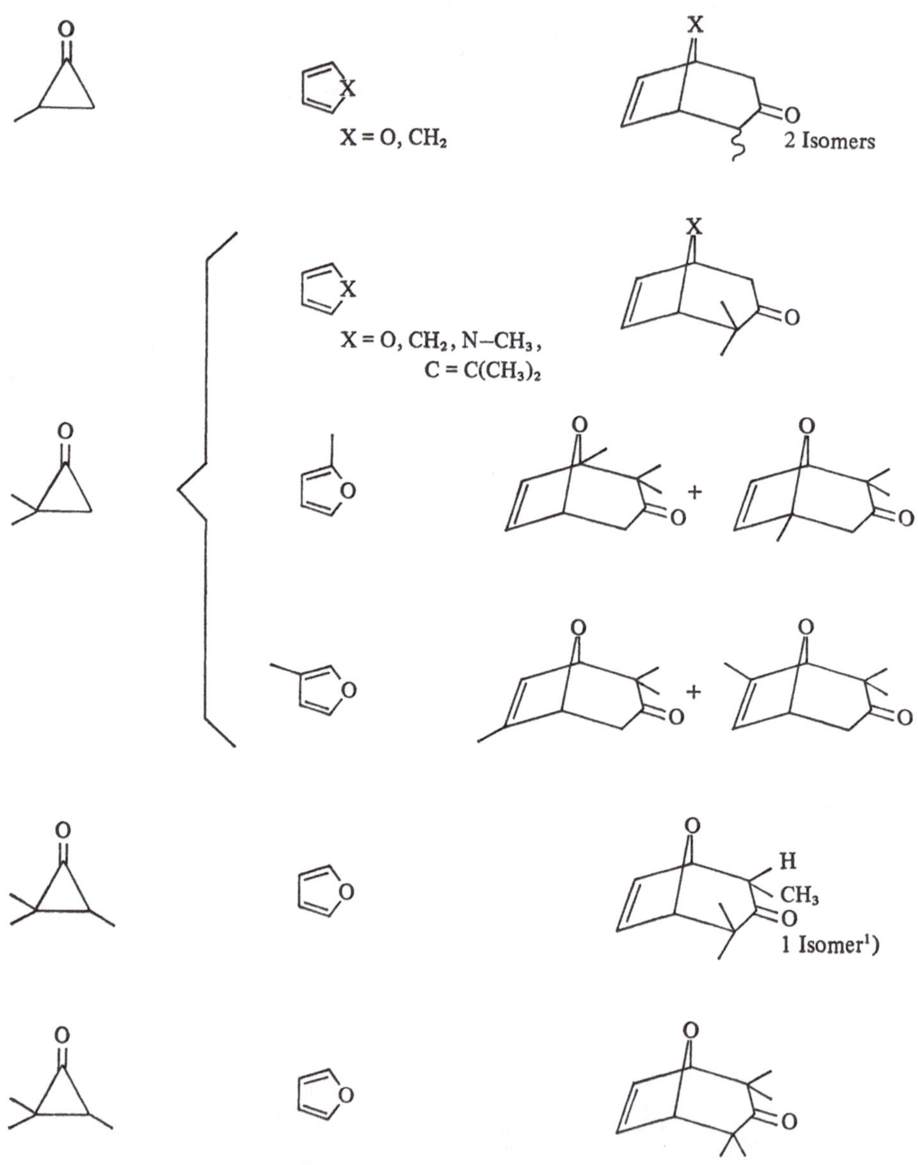

X = O, CH₂

2 Isomers

X = O, CH₂, N—CH₃, C = C(CH₃)₂

+

+

H
CH₃
O
1 Isomer[1)]

4.4.2. $3+2 \rightarrow 5$ Cycloadditions

The C_1–C_2 bond of cyclopropanones undergoes $3+2 \rightarrow 5$ cycloadditions with a variety of reactive two π electron systems by a thermally allowed process in which cyclopropanones function as nucleophilic 1,3-dipolar species. [118,122] Thus, 2,2-dimethylcyclopropanone reacts with excess chloral to give the adducts *151* and *152*. Other examples are given in Table 24.

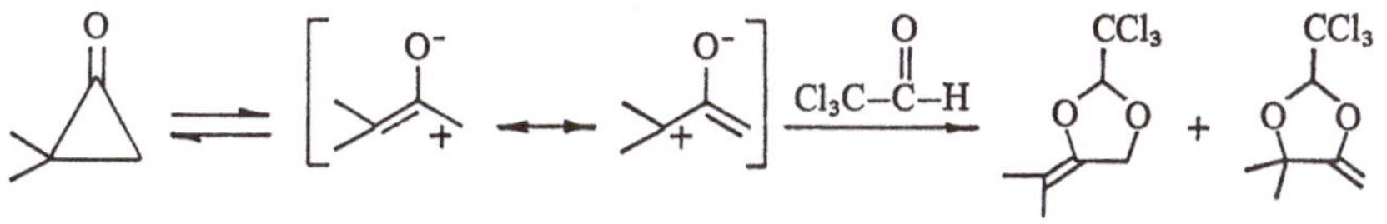

A mechanism involving two competing $3+2 \rightarrow 5$ cycloadditions has been suggested for the reaction between oxygen and tetramethylcyclopropanone (Scheme 33). [81] However, in view of the CIDNP effects observed by de Boer [115] upon exposure of 2,2-dimethylcyclopropanone to air, a radical process seems more likely.

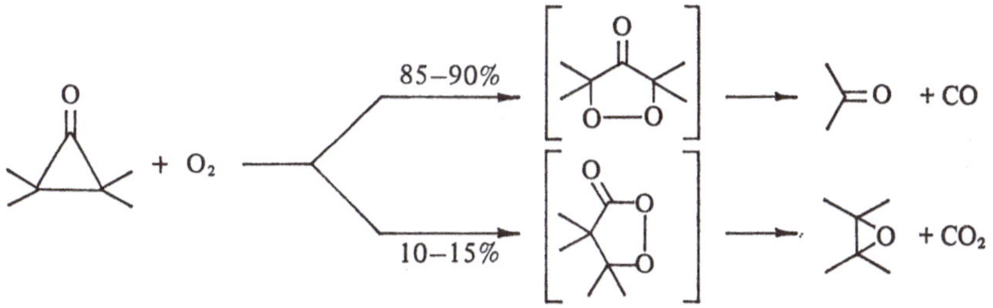

Scheme 33

4.4.3. $2+2 \rightarrow 4$ Cycloadditions

The third type of cycloaddition results from the reaction of cyclopropanones with activated olefins. For example, dimethylketene adds to methyl substituted cyclopropanones affording the spiro lactones *153a–c*. [96,118,122b] Similarly the *ortho* ester *154* is formed from 1,1-dimethoxyethylene and 2,2-dimethylcyclopropanone; *154* dimerizes to *155* upon standing. [118]

Table 24. 3+2→5 Cycloaddition reactions with 2,2-dimethylcyclopropanone [118)]

Dipolarophile	Addition products
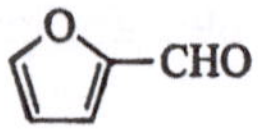	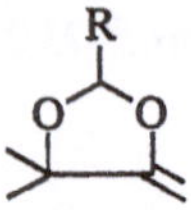 R = furfuryl (also 4 + 3→7 product)
Cl_3C-CHO	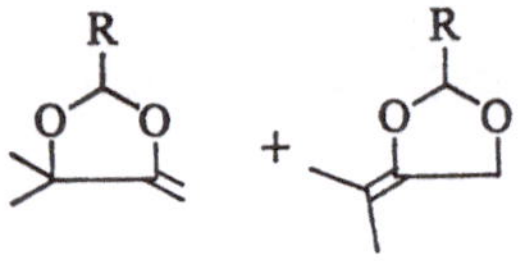 R = CCl_3
C_6H_5-CHO	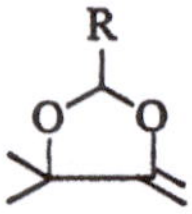 R = C_6H_5
CH_3CHO	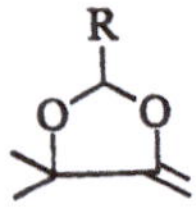 R = CH_3
SO_2	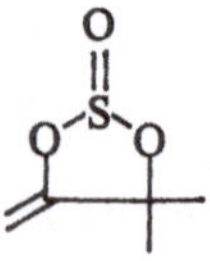

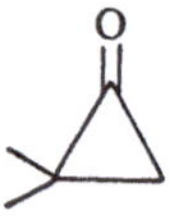

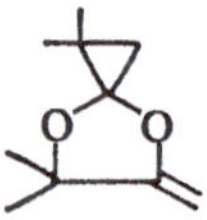

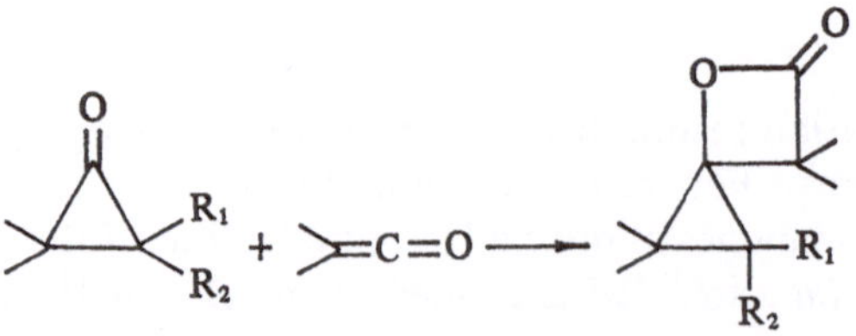

a, $R_1=R_2=H$
b, $R_1=H$; $R_2=CH_3$
c, $R_1=R_2=CH_3$

153

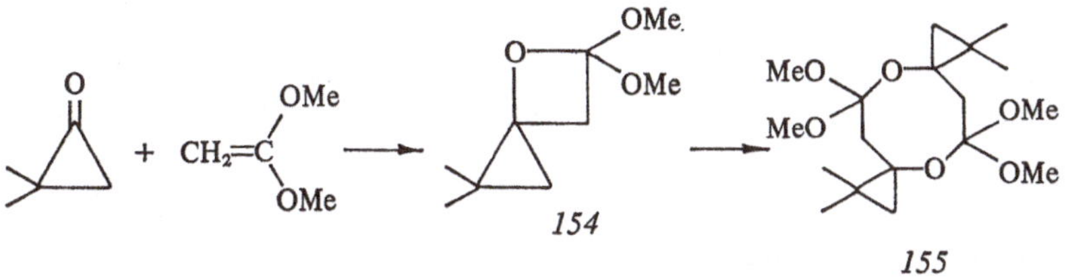

Since concerted $2+2\rightarrow4$ additions are forbidden by symmetry rules [123], the adducts *153* and *154* may be considered as further, albeit unusual, examples of a stepwise reaction involving an initial facile addition to the reactive carbonyl group (see Section 4.1.1—4.1.3).

4.5. Cyclopropanone Reductions

Upon treatment with lithium aluminum hydride, cyclopropanones are reduced to cyclopropanols, *e.g.*, *18→156*. [8] 1-Ethoxycyclopropyl alcohol and acetate *(7a)* yield cyclopropanol in a similar manner, presumably through a first stage decomposition to ketone. [4]

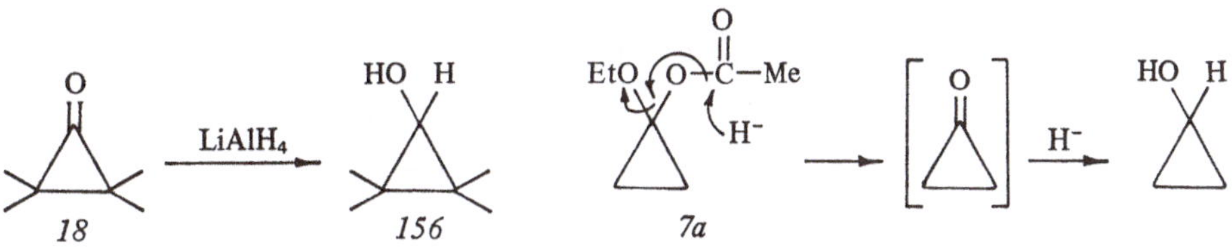

It has been reported that gem-dihalocyclopropanes, usually undergo electrochemical or metallic reduction with predominant retention of configuration as shown in Scheme 34. [124,126] The stereoselectivity of these reactions may result from the intermediacy of a cyclopropyl carbanion which is unusually stable to racemization compared to other alkyl carbanions. [125]

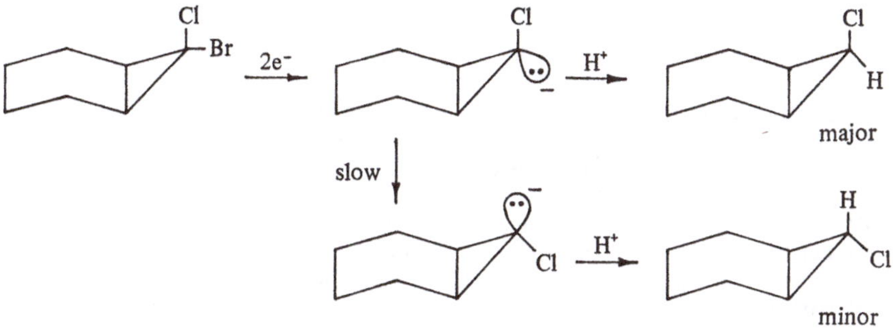

Scheme 34

151

H. H. Wasserman, G. M. Clark, and P. C. Turley

5. References

[1] Lipp, P., Buchkremer, J., Seeles, H.: Ann. *499*, 1 (1932).

[2] Turro, N. J., Hammond, W. B.: J. Am. Chem. Soc. *88*, 3672 (1966).

[3] Schaafsma, S. E., Steinberg, H., de Boer, T. J.: Rec. Trav. Chim. *85*, 1170 (1966).

[4] Wasserman, H. H., Clagett, D. C.: J. Am. Chem. Soc. *88*, 5368 (1966).

[5] Turro, N. J., Hammond, W. B.: Tetrahedron *24*, 6029 (1968).

[6] Wasserman, H. H., Baird, M. S.: Tetrahedron Letters *1970*, 1729.

[7] Wasserman, H. H., Clagett, D. C.: Tetrahedron Letters *1964*, 341.

[8] Turro, N. J., Hammond, W. B., Leermakers, P. A.: J. Am. Chem. Soc. *87*, 2774 (1965).

[9] a) Richey, Jr., H. G., Richey, J. M., Clagett, D. C.: J. Am. Chem. Soc. *86*, 3906 (1964);
b) Leermakers, P. A., Vesley, G. F., Turro, N. J., Neckers, D. C.: *86*, 4213 (1964).

[10] Turro, N. J., Hammond, W. B.: Tetrahedron *24*, 6017 (1968).

[11] a) Giusti, G., Morales, C., Feugeas, C.: Compt. Rend., Ser. C. *269*, 162 (1969);
b) Giusti, G., Morales, C.: Bull. Soc. Chim. France *1973*, 382.

[12] Fry, A. J., Scoggins, R.: Tetrahedron Letters *1972*, 4079.

[13] Pazos, J. F., Greene, F. D.: J. Am. Chem. Soc. *89*, 1030 (1967).

[14] van Tilborg, W. J. M.: Ph. D. Thesis, Univ. of Amsterdam, 1971.

[15] Turro, N. J., Hammond, W. B.: J. Am. Chem. Soc. *89*, 1028 (1967).

[16] Bakker, B. H.: Ph. D. Thesis, Univ. of Amsterdam, 1972.

[17] Dull, M. F., Abend, P. G.: J. Am. Chem. Soc. *81*, 2588 (1959).

[18] Wenkert, E., Mueller, R. A., Reardon, Jr., E. J., Sathe, S. S., Sharf, D. J., Tosi, G.: J. Am. Chem. Soc. *92*, 7428 (1970).

[19] a) Staudinger, H., Anthes, E., Pfenniger, F.: Ber. *49*, 1928 (1916). — Staudinger, H., Reber, T.: Helv. Chim. Acta. *4*, 3 (1921);
b) Kirmse, W.: *Ber. 93*, 2357 (1960).

[20] Kende, A. S.: Chem. Ind. (London) *1956*, 1053.

[21] Price, C. C., Vittimberga, J. S.: J. Org. Chem. *27*, 3736 (1962).

[22] a) Simmons, H. E., Smith, R. D.: J. Am. Chem. Soc. *81*, 4256 (1959);
b) Simmons, H. E., Blanchard, E. P., Smith, R. D.: J. Am. Chem. Soc. *86*, 1347 (1964).

[23] It is possible that a dialkylzinc-methylene iodide reagent may further improve the yield as in the case with cyclopropanes prepared from vinyl ethers (Furakawa, J., Kawabata, N., Nishimura, J.: *Tetrahedron 24*, 53 (1968)).

[24] Grewe, R., Struve, A.: Ber. *96*, 2819 (1963).

[25] McElvain, S. M., Weyna, P. L.: J. Am. Chem. Soc. *81*, 2579 (1959).

[26] a) Schöllkopf, U., Lehmann, G. J.: Tetrahedron Letters *1962*, 165;
b) Schöllkopf, U., Lehmann, G. J., Paust, J., Hartl, H.-D.: Ber. *97*, 1527 (1964);
c) Schöllkopf, U., Woerner, F. P., Wiskott, E.: Ber. *99*, 806 (1966).

[27] Schöllkopf, U., Gorth, H.: Ann. *709*, 97 (1967).

[28] Schöllkopf, U., Wiskott, E.: Ann. *694*, 44 (1966).

[29] Moss, R. A., Pilkiewicz, F. G.: Synthesis *1973*, 209 B.

[30] Schöllkopf, U., Paust, J.: Angew. Chem. *75*, 670 (1963); Chem. Ber. *98*, 2221 (1965).

[31] Hostettler, H. U.: Helv. Chim. Acta *49*, 2417 (1966).

[32] Turro, N. J.: Acc. Chem. Res. *2*, 25 (1969).

[33] Weinshenker, N. M., Greene, F. D.: J. Am. Chem. Soc. *90*, 506 (1968).

[34] Barber, L. L., Chapman, O. L., Lassila, J. D.: J. Am. Chem. Soc. *91*, 3664 (1969).

35) Kropp, P. J., Erman, W. F.: J. Am. Chem. Soc. *85*, 2456 (1963).

36) Koch, T. H., Sluski, R. J.: Tetrahedron Letters, *1970*, 2391.

37) Agosta, W. C., Smith, III, A. B.: Tetrahedron Letters, *1969*, 4517.

38) Wagner, P. J., Bucheck, D. J.: J. Am. Chem. Soc. *91*, 5090 (1969) and references therein.

39) Cookson, R. C., Nye, N. J.: Proc. Chem. Soc, *1963*, 129.

40) O'Brien, J. P., Rachlin, A. I., Teitel, S.: J. Med. Chem. *12*, 1112 (1969).

41) Feugeas, C., Galy, J. P.: Compt. Rend., Ser. C. *270*, 2157 (1970).

42) Doering, W. von E., Hoffman, A. K.: J. Am. Chem. Soc. *76*, 6162 (1954).

43) Parham, W. E., Schweizer, E. E.: Org. Reactions *13*, 55 (1969).

44) Makosza, M., Wawrzyniewicz, M.: Tetrahedron Letters *1969*, 4659.

45) a) Weyerstahl, P., Blume, G., Muller, C.: Tetrahedron Letters *1971*, 3869;
b) Weyerstahl, P., Mathias, R., Blume, G.: Tetrahedron Letters *1973*, 611..

46) Chau, L. V., Schloss, M.: Synthesis *1973*, 112.

47) Groves, J. T., Ma, K. W.: Abst. A. C. S. Meeting, Chicago, Aug. 1973.

48) Shields, T. C., Gardner, P. D.: J. Am. Chem. Soc. *89*, 5425 (1967).

49) a) Loftfield, R. B.: J. Am. Chem. Soc. *72*, 632 (1950); *73*, 4707 (1951);
b) For a review of the Favorskii rearrangement, see Kende, A. S.: Org. Reactions *11*, 261 (1960).

50) Breslow, R., Eicher, T., Krebs, A., Peterson, R. A., Posner, J.: J. Am. Chem. Soc. *87*, 1320 (1965).

51) Camp, R. L., Greene, F. D.: J. Am. Chem. Soc. *90*, 7349 (1968).

52) a) Szmuszkovicz, J., Cerda, E., Grostic, M. F., Zieserl, Jr., J. F.: Tetrahedron Letters *1967*, 3969;
b) Szmuszkovicz, J., Duchamp, D. J., Cerda, E., Chidester, C. G.: Tetrahedron Letters *1969*, 1309.

53) Quast, H., Schmitt, E., Frank, R.: Angew. Chem. Intern. Ed. Engl. *10*, 651 (1971).

54) Quast, H., Schmitt, E.: Ber. *103*, 1234 (1970).

55) a) Crandall, J. K., Machleder, W. H.: J. Am. Chem. Soc. *90*, 7347 (1968);
b) Crandall, J. K., Machleder, W. H., Thomas, M. J.: J. Am. Chem. Soc. *90*, 7346 (1968);
c) Crandall, J. K., Machleder, W. H., Sojka, S. A.: J. Org. Chem. *38*, 1149 (1973).

56) Hoffmann, H. M. R., Smithers, R. H.: Angew. Chem. Intern. Ed. Engl. *9*, 71 (1970).

57) Skorianetz, W., Schulte-Elte, K. H., Ohloff, G.: Helv. Chim. Acta *54*, 1913 (1971).

58) Harada, N., Suzuki, S., Uda, H., Ueno, H.: J. Am. Chem. Soc. *94*, 1777 (1972).

59) a) Deyrup, J. A., Greenwald, R. B.: Tetrahedron Letters *1966*, 5091;
b) Sheehan, J. C., Nafissi-V, M. M.: J. Am. Chem. Soc. *91*, 4596 (1969). Other workers were not able to duplicate some results;
c) Talaty, E. R., DuPuy, Jr., A. E., Johnson, C. K., Pirotte, T. P., Fletcher, W. A., Thompson. R. E.: Tetrahedron Letters *1970*, 4435.

60) Quast, H., Risler, W.: Angew. Chem. Intern. Ed. Engl. *85*, 411 (1973).

61) Baucom, K. B., Butler, G. B.: J. Org. Chem. *37*, 1730 (1972).

62) Breslow, R., Oda, M.: J. Am. Chem. Soc. *94*, 4787 (1972).

63) Pochan, J. M., Baldwin, J. E., Flygare, W. H.: J. Am. Chem. Soc. *91*, 1896 (1969); *90*, 1072 (1968).

64) Sclove, D. B., Pazos, J. F., Camp. R. L., Greene, F. D.: J. Am. Chem. Soc. *92*, 7488 (1970).

65) Peter, R., Dreizler, H.: Z. Naturforsch. *20a*, 301 (1965).

H. H. Wasserman, G. M. Clark, and P. C. Turley

66) Shoolery, J. N., Sharbaugh, A. H.: Phys. Rev. *82*, 95 (1951).
67) a) Rao, C. N. R.: Chemical applications of infrared spectroscopy. New York: Academic Press 1963.
   b) Bellamy L. J.: The infrared spectra of complex molecules, 2nd ed. New York: John Wiley and Sons, Inc. 1958.
68) van Tilborg, W. J. M.: Tetrahedron Letters *1973*, 523.
69) a) Brauman, J. I., Laurie, V. W.: Tetrahedron *24*, 2595 (1968);
   b) Davis, R. E., Grosse, D. J.: Tetrahedron *26*, 1171 (1970);
   c) Galabov, B., Simov, D.: Chem. Phys. Letters *5*, 549 (1970).
70) Scott, A. I.: Interpretation of the ultraviolet spectra of natural products. New York: Macmillan 1964.
71) Rao, C. N. R., Goldman, G. K., Ramachandran, J.: J. Indian Inst. Sci.*43*,: 10 (1961). — Rao, C. N. R.: Ultraviolet and visible spectroscopy, p. 19. London: Butterworth 1961.
72) Jackman, L. M., Sternhill, S.: Application of nuclear magnetic resonance spectroscopy in organic chemistry, 2nd ed. Oxford: Pergamon Press 1969.
73) Muller, N., Pritchard, D. E.: J. Chem. Phys. *31*, 768 (1959).
74) Jongejan, E., van Tilborg, W. J. M., Dusseau, C. H. V., Steinberg, H., deBoer, T. J.: Tetrahedron Letters *1972*, 2359.
75) Thomas, T. F., Rodriguez, H. J.: J. Am. Chem. Soc. *93*, 5918 (1971).
76) Wasserman, H. H., Cochoy, R. C., Baird, M. S.: J. Am. Chem. Soc. *91*, 2375 (1969).
77) Bodor, N., Dewar, M. J. S., Harget, A., Haselbach, E.: J. Am. Chem. Soc. *92* 3854 (1970).
78) Liberles, A., Kang, S., Greenberg, A.: J. Org. Chem. *38*, 1922 (1973).
79) Edelson, S. S., Turro, N. J.: J. Am. Chem. Soc. *92*, 2770 (1970).
80) Schaafsma, S. E., Steinberg, H., de Boer, T. J.: Rec. Trav. Chim. *86*, 561 (1967)
81) Turro, N. J., Leermakers, P. A., Wilson, H. R., Neckers, D. C., Byers, G. W. Vesley, G. F.: J. Am. Chem. Soc. *87*, 2613 (1965).
82) Wasserman, H. H., Clark, G. M.: unpublished results.
83) van Tilborg, W. J. M., Schaafsma, S. E., Steinberg, H., deBoer, T. J.: Rec. Trav. Chim. *86*, 419 (1967).
84) Wasserman, H. H., Adickes,.: H. W.: unpublished results.
85) a) Conia, J. M., Dennis, J. M.: Tetrahedron Letters *1971*, 2845;
   b) Heine, H. G.: Chem. Ber. *104*, 2869 (1971).
86) van Tilborg, W. J. M., Schaafsma S. E., Steinberg, H., de Boer, T. J.: Rec. Trav. Chim. *86*, 417 (1967).
87) a) Wasserman, H. H., Adickes, H. W., de Ochoa, O. E.: J. Am. Chem. Soc. *93*, 5586 (1971);
   b) Wasserman, H. H., Glazer, E.: unpublished results.
88) Roberts, J. D., Chambers, V. C.: J. Am. Chem. Soc. *73*, 5034 (1951).
89) Wasserman, H. H., Baird, M. S.: Tetrahedron Letters *1971*, 3721.
90) Turro, N. J., Hammond, W. B.: Tetrahedron Letters *1967*, 3085.
91) Howell, B. A., Jewett, J. G.: J. Am. Chem. Soc. *93*, 798 (1971).
92) van Tilborg, W. J. M., Dooycevaard, G., Steinberg, H., de Boer, T.J.,: Tetrahedron Letters *1972*, 1677.
93) Wasserman, H. H., Haveaux, B., Thyes, M.: unpublished results.
94) Dalton, J. C., Turro, N. J.: Ann. Rev. Phys. Chem. *21*, 499 (1970).
95) Dalton, J. C., Pond, D. M., Weiss, D. S., Lewis, F. D., Turro, N. J.: J. Am. Chem. Soo. *92*, 2564 (1970).
96) Turro, N. J., Gagosian, R. B.: Chem. Commun. *1969*, 949; J. Am. Chem. Soc. *92*, 2036 (1970).

97) Lipp, P., Koster, R.: Ber. *64*, 2823 (1931).
98) a) Conia, J. M., Salaun, J.: Bull. Soc. Chim. France *1964*, 1957;
    b) Salaun, J., Conia, J.-M.: Bull. Soc. Chim. France *1968*, 3730.
99) Semenow, D. A., Cox, E. F., Roberts, J. D.: J. Am. Chem. Soc. *78*, 3221 (1956).
100) Wasserman, H. H., Hearn, M., Glazer, E.: unpublished results.
101) a) Turro, N. J., Gagosian, R. B., Rappe, C., Knutsson, L.: Chem. Commun. *1969*, 270.
    b) Rappe, C., Knutsson, L., Turro, N. J., Gagosian, R. B.: J. Am. Chem. Soc. *92*, 2032 (1970).
102) Wharton, P. S., Fritzberg, A. R.: J. Org. Chem. *37*, 1899 (1972).
103) Hammond, W. B., Turro, N. J.: J. Am. Chem. Soc. *88*, 2880 (1966).
104) Turro, N. J., Hammond, W. B., Leermakers, P. A., Thomas, H. T.: Chem. Ind. (London) *1965*, 990.
105) Giusti, G.: Compt. Rend. Ser. C. *273*, 257 (1971); Bull. Soc. Chim. France *1972*, 4335.
106) For a general review, see Hunig, S.: Angew. Chem. Intern. Ed. Engl. *3*, 548 (1964).
107) Giusti, G., Vincent, E. J.: Bull. Soc. Chim. France 803 (1973).
108) Quast, H., Frank, R., Schmitt, E.: Angew. Chem. Intern. Ed. Engl. *11*, 329 (1972).
109) Dusseau, C. H. V., Schaafsma, S. E., de Boer, T. J.: Rec. Trav. Chim. *89*, 535 (1970).
110) Seebach, D., Braun, M.: Angew. Chem. Intern. Ed. Engl. *11*, 49 (1972).
111) Ledlie, D. B.: J. Org. Chem. *37*, 1439 (1972).
112) a) Schaafsma, S. E., Steinberg, H., de Boer, T. J.: Rec. Trav. Chim. *85*, 70 (1966);
    b) Schaafsma, S. E., Steinberg, H., de Boer, T. J.: Rec. Trav. Chim. *85*, 73 (1966).
113) Schaafsma, S. E., Molenaar, E. J. F., Steinberg, H., de Boer, T. J.: Rec. Trav. Chim. *86*, 1301 (1967).
114) Schaafsma, S. E., Jorritsma, R., Steinberg, H., de Boer, T. J.: Tetrahedron Letters *1973*, 827.
115) a) Bakker, B. H., Schilder, G. J. A., Bok, T. R., Steinberg, H., de Boer, T. J.: Tetrahedron *29*, 93 (1973);
    b) de Boer, T. J.: Angew. Chem. Intern. Ed. Engl. *11*, 321 (1972).
116) a) Gibson, D. H., DePuy, C. H.: Tetrahedron Letters *1969*, 2203;
    b) Priddy, D. B., Reusch, W.: Tetrahedron Letters *1970*, 2637.
117) DePuy, C. H., Jones, H. L., Gibson, D. H.: J. Am. Chem. Soc. *90*, 306 (1968).
118) Turro, N. J., Edelson, S. S., Williams, J. R., Darling, T. R., Hammond, W. B.: J. Am. Chem. Soc. *91*, 2283 (1969) and references therein.
119) Fort, A. W.: J. Am. Chem. Soc. *84*, 4979 (1962).
120) Hoffmann, H. M. R., Clemens, K. E., Smithers, R. H.: J. Am. Chem. Soc. *94*, 3940 (1972).
121) Turro, N. J., Edelson, S. S.: J. Am. Chem. Soc. *90*, 4499 (1968).
122) a) Turro, N. J., Edelson, S. S., Williams, J. R., Darling, T. R.: J. Am. Chem. Soc. *90*, 1926 (1968);
    b) Turro, N. J., Edelson, S. S., Gagosian, R. B.: J. Org. Chem. *35*, 2058 (1970).
123) Woodward, R. B., Hoffmann, R.: The conservation of orbital symmetry. New York: Academic Press 1970.
124) Erickson, R. E., Annino, R., Scanlon, M. D., Zon, G.: J. Am. Chem. Soc. *91*, 1767 (1969).

H. H. Wasserman, G. M. Clark, and P. C. Turley

125) Walborsky, H. M., Impastato, F. J., Young, A. E.: J. Am. Chem. Soc. *86*
3283 (1964).
126) Fry, A. J.: Fortschritte *34*, 1 (9172).
127) DeMore, W. B., Pritchard, H. O., Davidson, N.: J. Am. Chem. Soc. *81*, 5874
(1959).
128) Haller, I., Srinivasan, R.: J. Am. Chem. Soc. *87*, 1144 (1965); Can. J. Chem. *43*,
3165 (1965).
129) Robinson, G. C.: Tetrahedron Letters *1965*, 1749.
130) Cochoy, R. C.: Ph. D. Thesis, Yale University, 1969.
131) Zecher, D. C., West, R.: J. Am. Chem. Soc. *89*, 153 (1967).
132) Turro, N. J., Hammond, W. B.: J. Am. Chem. Soc. *87*, 3258 (1965).
133) van Tilborg, W. J. M., Steinberg, H., de Boer, T. J.: Syn. Commun. *3*, 189
(1973).

Received September 28, 1973

J. H. van't Hoff: Imagination in Science

Translated into English with notes and a general
introduction by G. F. Springer
1 portrait. VI, 18 pages. 1967 (Molecular Biology,
Biochemistry and Biophysics, Vol. 1)
DM 6,60; US $2.60
ISBN 3-540-03933-3

This small volume comprises the inaugural lecture of
the first Nobel Laureate in chemistry, Jacobus Henricus
van't Hoff.
Because of its inspirational nature, it has been selected
as the initial volume in the new Springer-Verlag series
"Molecular Biology, Biochemistry and Biophysics". We
are also convinced of its very special educational merits
for the training of young chemists.
The monograph shows the two possible successful
approaches to science, the purely empirical one and
the one based on ideas but controlled by facts. Each
is represented by its protagonists, the former by Kolbe,
the latter by van't Hoff. Most important for a scientist
as well as a beginner, principles of productive scientific
research and its preconditions, as well as the way of
scientific thinking, are developed here in a clear, uni-
versally valid fashion.

W. Bähr; H. Theobald: Organische Stereochemie Begriffe und Definitionen

XV, 122 pages. 1973 (Heidelberger Taschenbücher,
Bd. 131) DM 16,80; US $6.50
IBSN 3-540-06339-0

Over the years many new words have been coined in
organic stereochemistry, and some old ones have been
redefined. This compilation of 89 main concepts with
300 alphabetically arranged definitions gives priority to
static stereochemistry. It offers a source of rapid information for working scientists—biochemists, molecular
biologists and physicians—as well as for students and
teachers of chemistry.

Prices are subject to change without notice

Springer-Verlag
Berlin Heidelberg New York

STRUCTURE AND BONDING

Editors:
J. D. Dunitz; P. Hemmerich;
J. A. Ibers; C. K. Jørgensen;
J. B. Neilands; D. Reinen;
R. J. P. Williams

Prices are subject to change without notice

 **Springer-Verlag
Berlin
Heidelberg
New York**